D1481872

Linear
Free
Energy
Relationships

Linear
Free
Energy
Relationships

P. R. WELLS

*Department of Chemistry, University of Queensland,
St. Lucia, Brisbane, Australia*

LONDON · NEW YORK
1968

INDIANA UNIVERSITY
LIBRARY
NORTHWEST

ACADEMIC PRESS INC. (LONDON) LTD.
Berkeley Square House, Berkeley Square
London, W.1

U.S. Edition published by
ACADEMIC PRESS INC.
111 Fifth Avenue, New York, New York 10003

PRINTED IN GREAT BRITAIN BY
J. W. ARROWSMITH LTD., WINTERSTOKE ROAD, BRISTOL 3

Preface

The correlation of relative reactivities in terms of relative free energy changes has provided quantitative bases in terms of which observations may be discussed and theories of reactivity developed. This approach has become a well defined section of physical organic chemistry, and is being applied with advantage in the fields of inorganic and biochemical reactivity studies.

In this monograph the well established relationships are critically surveyed under the major headings of substrate structure, reagent structure and reaction medium. Substituent effects are examined in detail for various substrate types and theoretical treatments are discussed in terms of the implications of the linear free energy relationships. The correlation of solvent effects upon solvolysis rates and upon acid strength are described and related to the concept of acidity functions. The relationships between quantities derived from reactivity and from spectroscopic studies are also examined.

To provide an instructive and informative text for advanced students and research workers, the subject has been approached critically and with some scepticism. Nevertheless an attempt has been made to extract the maximum information provided by the relationships in the discussions of the parameters obtained. Several tables of directly obtained empirical parameters and certain derived quantities are given.

A large part of the preparation of this manuscript was carried out during stimulating visits to the Departments of Chemistry of the University of California, Irvine, and the State University of New York, Stony Brook, whose generous hospitality is acknowledged. Of many helpful discussions with faculty members and graduate students I am particularly grateful to Professors Robert W. Taft and Paul C. Lauterbur.

Brisbane   P. R. WELLS
April, 1968

Contents

1

Introduction

During the past three decades examination of the rapidly accumulating data on reaction rates and equilibria has led to the formulation of several empirical correlations. The general form of these correlations is a linear relationship between the logarithms of the rate or equilibrium constants for one reaction series (k_i^B) and those for a second reaction series (k_i^A) subjected to the same variations in reactant structure or reaction conditions. The relationship may be expressed by equation 1.1

$$\log k_i^B = m \log k_i^A + c \qquad (1.1)$$

where m is the slope and c the intercept of the straight line obtained.

A set of rate or equilibrium constants corresponding to essentially the same reaction of a series of substrates, of a given substrate with a series of reagents or of a single reaction under a variety of environmental conditions, e.g. solvent changes, is termed a reaction series.

The logarithm of an equilibrium constant (K) is proportional to the standard free energy change (ΔF^0) accompanying the reaction

$$\log K = -\Delta F^0/2\cdot303RT \qquad (1.2)$$

and, according to the transition-state theory, a specific rate constant (k) can be expressed in terms of a standard free energy of activation ($\Delta F^\ddagger$)

$$\log k = \log(RT/Nh) - \Delta F^\ddagger/2\cdot303RT \qquad (1.3)$$

Combining equations 1.1, 1.2 and 1.3 one obtains

$$\Delta F^B = n\Delta F^A + d \qquad (1.4)$$

where the relationship between n and m and between d and c depends upon whether the reactivity comparison is expressed in terms of equilibrium constants, or rate constants or both.

When either or both rate and equilibrium data are correlated, the symbols k and ΔF will be taken to refer to either type of information.

The empirical correlation of reactivity changes by means of equation 1.1 is thus equivalent to a *linear free energy relationship* (LFER), i.e. equation 1.4.

The formal similarity of equations 1.2 and 1.3 provides an explanation for the correlation of both equilibrium and rate data. Since the former are independent but the latter completely dependent upon the reaction path between reactants and products, their simultaneous correlation might have been unexpected.

Although Ostwald (1885) was among the earliest to recognize regularities within a reaction series, the strength of aliphatic acids, the first comparisons of several series of acid strengths in terms of equation 1.1 was that of Hixon and Johns (1927). The first LFER to be recognized, the Brønsted equation, was reported by Brønsted and Pederson (1924) in relation to the relative effectiveness of acids and bases as catalysts. Correlations of substituent effects rapidly followed, particularly for aromatic substrates. Kindler (1928) found a relationship between the alkaline hydrolysis rates of substituted ethyl benzoates and ethyl cinnamates, Hammett and Pfluger (1933) correlated the reaction of trimethylamine with substituted methyl benzoates and the strengths of the corresponding benzoic acids. These extensions of the Hixon–Johns' approach soon found general expression as the Hammett (1937) equation. At the same time Nathan and Watson (1933) and Dippy and Watson (1936) were formulating substituent effects on reactivity in terms of substituent group dipole moments. These correlations are the forerunners of several correlations of relative reactivity with physical properties.

Apart from qualitative statements concerning polarity and some correlation of acid strengths with dielectric constant, few correlations of the influence of the reaction medium on relative reactivity were apparent until the development of the Grunwald–Winstein equation (1948). Its original limited form has been extended into a general treatment of solvent effects, and deviations from the original equation have led to the formulation of equations correlating changes in structure of the reagent.

Of all possible relationships between observable quantities the linear form, equation 1.1, is the most easily recognized. This is particularly the case when the data are examined graphically. Although it is now the common practice to use statistical methods in the study of correlations, the first indication of the existence of a correlation and its qualitative assessment usually results from a plot of $\log k_i^B$ versus $\log k_i^A$.

Reaction series that follow equation 1.1 perfectly are extremely rare. Both k_i^B and k_i^A will contain random experimental errors leading to random deviations from the correlation. These are often not obvious, particularly when the number of data correlated is small. Statistical methods can be used to reveal deviations of this type and identify the members of the reaction series responsible. An improved correlation will result from the omission of these numbers. More frequently, one has other information and suspicions of deviating members and omissions from the correlation are made on these bases.

Having decided, either intuitively or by graphical examination, that $\log k_i^B$ may be more or less well correlated with $\log k_i^A$ by means of equation 1.1, statistical methods can be used to give the best estimates of m and c. This best estimate is interpreted as the values of m and c that minimize the sum of the squares of the deviations of the observed values of $\log k_i^B$ from the values given by equation 1.1. Equation 1.1 derived in this way is termed the regression line of $\log k_i^B$ on $\log k_i^A$. It should be noted that whatever the form of the association between the two observables, only the best linear relationship has been determined. An assessment of the validity of this assumed relationship can also be obtained from the statistical analysis. The magnitude of the correlation coefficient, r, may be taken as a measure of the degree to which the association approaches a linear relationship. Conventionally one describes correlations having $r > 0.99$ as excellent, $r > 0.95$ as satisfactory and $r > 0.90$ as fair.

The correlation of multiple variations, e.g. by equation 1.5

$$\log k_i^B = m \log k_i^A + m' \log k_i^{A'} + c \qquad (1.5)$$

is more difficult to observe graphically but is readily handled by statistical methods. It is important to realize that equation 1.5 cannot correlate the data less well than equation 1.1. If the term $m' \log k_i^{A'}$ is regarded simply as a correcting factor for equation 1.1, then equation 1.5 must give a better correlation unless $\log k_i^{A'}$ bears absolutely no relationship to the deviations from equation 1.1. In this extreme case statistical analysis will give $m' = 0$ making equations 1.1 and 1.5 identical. Statistical procedures are available with which to determine whether equation 1.5 is a significant improvement on equation 1.1, and whether there is a "real" relationship between $\log k_i^B$ and both $\log k_i^A$ and $\log k_i^{A'}$.

A meaningful relationship must encompass at least four $\log k_i^B$ values for each series and be applicable to several reaction series. Two $\log k_i^B / \log k_i^A$ pairs must lie on a straight line, and even three pairs that are reasonably well correlated may be a chemically trivial result. A relationship applicable to a very small number of reaction series will have only limited value. Thus although there are relatively few LFER of general use to be described, there may be several special sets of circumstances for which equations of the form of equation 1.1 and 1.5 are applicable.

If a member of one reaction series is chosen as a reference, e.g. k_0^A, then, if equation 1.1 holds, the corresponding member of the compared series, k_0^B, is given by

$$\log k_0^B = m \log k_0^A + c \qquad (1.6)$$

and equation 1.1 becomes

$$\log(k_i^B / k_0^B) = m \log(k_i^A / k_0^A) \qquad (1.7)$$

Equation 1.7 is a comparison of relative reactivities, as in reality is equation 1.1. The right-hand side is composed of two factors: $\log(k_i^A/k_0^A)$, which is independent of the reaction series B and may be used as a measure of the structural or environmental change; and m, which is independent of the magnitude of the change, but depends upon the type of change and the two reaction series compared. If these factors are suitably defined and symbolized, for example by

$$X_i = \log(k_i^A/k_0^A)$$

$$m = G_X^{AB}$$

the general form of the two-parameter LFER, equation 1.8, is obtained

$$\log(k_i^B/k_0^B) = G_X^{AB} \cdot X_i \tag{1.8}$$

The general form of the four-parameter LFER, equation 1.9, arises similarly from equation 1.5

$$\log(k_i^B/k_0^B) = G_X^{AB} \cdot X_i + G_Y^{A'B} \cdot Y_i \tag{1.9}$$

The most valuable feature of the LFER's is that they permit one to recognize regular patterns of chemical behaviour and readily to observe deviations from these patterns. A quantitative expression of "normal" response of a system to variations in reactant structure and reaction medium is obtained. Established qualitative concepts become quantitative and provide the necessary stepping stones for the development of more precise theories. "Abnormal" behaviour can be identified only when the "normal" pattern has been determined. The incursion of factors previously not recognized and major changes in reaction mechanism are indicated by deviations from well established relationships.

Of practical value is the prediction of reactivity changes. In some cases this can be made with considerable confidence. A further economy arises in the storage of data. Ten X_i parameters and ten G_X^{AB} parameters correspond to 100 values of $\log(k_i^B/k_0^B)$.

Although the LFERs are completely empirical the search for correlations may be directed by some theoretical concept. Since the result of systematic variations in one reaction series are quantitatively proportional to the corresponding changes in another series, it may be inferred that the reaction variable in question operates in each series in essentially the same manner. Deviations will be observed if another variable is operative or if the single variable has some effect specific for one reaction series or individual members of a series. The LFERs themselves do not *identify* the variable in question. This can sometimes be inferred from the type of reaction series involved and various additional information concerning individual reactions. However, the mere existence of LFER's places certain demands upon theories of structural and environmental effects on reactivity.

Suppose the standard free energy change (ΔF) for a reaction can be expressed as a function of a number of variables, $x, y \ldots$. Changes in ΔF with changes in the variables will be given by

$$d\Delta F = (\partial \Delta F/\partial x)\,dx + (\partial \Delta F/\partial y)\,dy + \cdots \tag{1.10}$$

For a finite change in the variable x from some standard value x_0 to the value x_i while the remaining variables are held constant,

$$\Delta F_i - \Delta F_0 = (\partial \Delta F/\partial x)(x_i - x_0) \tag{1.11}$$

provided that $(\partial \Delta F/\partial x)$ remains constant throughout the range of variation of x, i.e. $(\partial^2 \Delta F/\partial x^2)$ etc. $= 0$.

If equation 1.8 holds for two reaction series, A and B, corresponding to changes in a single variable, x, then

$$\Delta F_i^B - \Delta F_0^B = \frac{(\partial \Delta F^B/\partial x)}{(\partial \Delta F^A/\partial x)} \cdot (\Delta F_i^A - \Delta F_0^A) \tag{1.12}$$

Equation 1.8 is a two-parameter LFER that, by means of equations 1.2 and 1.3, may be expressed in the form

$$\log(k_i^B/k_0^B) = \frac{T_A(\partial \Delta F^B/\partial x)}{T_B(\partial \Delta F^A/\partial x)} \cdot \log(k_i^A/k_0^A) \tag{1.13}$$

and the two parameters are defined by

$$X_i = \log(k_i^A/k_0^A) = -(\Delta F_i^A - \Delta F_0^A)/2\cdot303RT_A \tag{1.14}$$

(independent of reaction series B)

$$\text{and} \quad G_X^{AB} = \frac{T_A(\partial \Delta F^B/\partial x)}{T_B(\partial \Delta F^A/\partial x)} \tag{1.15}$$

(independent of the magnitude of x)

Two conditions for the observation of a LFER are thus (i) a single variable is changed, and (ii) $(\partial \Delta F/\partial x)$ is independent of x throughout the range x_0 to x_i. The relaxation of these conditions may lead to the failure of equation 1.7 or deviations from it. If several variables are changed equation 1.7 may still be applicable. This could arise if there are simultaneous variations in y, z etc. but these were proportional to the variations in x, i.e.

$$dy = a_{xy}\,dx; \quad dz = a_{xz}\,dx \text{ etc.}$$

where a_{xy} etc. are independent of x and y. In this case

$$\Delta F_i^B - \Delta F_0^B = \left[\frac{(\partial \Delta F^B)}{\partial x} + a_{xy}^B \frac{(\partial \Delta F^B)}{\partial y} + \cdots \right] (x_i - x_0) \qquad (1.16)$$

so that

$$\log(k_i^B/k_0^B) = \frac{T_A}{T_B} \frac{[(\partial \Delta F^B/\partial x) + a_{xy}^B (\partial \Delta F^B/\partial y) + \cdots]}{[(\partial \Delta F^A/\partial x) + a_{xy}^A (\partial \Delta F^A/\partial y) + \cdots]} \log(k_i^A/k_0^A) \qquad (1.17)$$

which, from the viewpoint of observing a LFER, is indistinguishable from equation 1.13.

When changes in ΔF^B arise from two variables, x and y, then it may be possible to correlate these changes by means of two reference series A and A', corresponding, respectively, to changes in x alone and y alone.

$$\Delta F_i^A - \Delta F_0^A = (\partial \Delta F^A/\partial x)_s (x_i - x_0) \qquad (1.18)$$

$$\Delta F_i^{A'} - \Delta F_0^{A'} = (\partial \Delta F^{A'}/\partial y)_{s'} (y_i - y_0) \qquad (1.19)$$

In equations 1.18 and 1.19 the reference series correspond to constant values y_s and $x_{s'}$, respectively.

Since, in general, $(\partial \Delta F^B/\partial x)$ will be a function of y, and $(\partial \Delta F^B/\partial y)$ will be a function of x

$$\Delta F_i^B - \Delta F_0^B = (\partial \Delta F^B/\partial x)y_0(x_i - x_0) + (\partial \Delta F^B/\partial y)x_0(y_i - y_0)$$
$$+ 2(\partial^2 \Delta F^B/\partial x \partial y)(x_i - x_0)(y_i - y_0) \qquad (1.20)$$

and hence

$$\log(k_i^B/k_0^B)$$

$$= \frac{T_A}{T_B} \frac{(\partial \Delta F^B/\partial x)y_0}{(\partial \Delta F^A/\partial x)_s} \cdot \log(k_i^A/k_0^A) + \frac{T_A}{T_B} \frac{(\partial \Delta F^B/\partial z)x_0}{(\partial \Delta F^{A'}/\partial y)_{s'}} \cdot \log(k_i^{A'}/k_0^{A'})$$

$$- \frac{2 \cdot 303RT_A \cdot T_{A'}}{T_B} \frac{(\partial^2 \Delta F^B/\partial x \partial y)}{(\partial \Delta F^A/\partial x)(\partial \Delta F^{A'}/\partial y)} \cdot \log(k_i^A/k_0^A) \cdot \log(k_i^A/k_0^{A'})$$

$$= G_X^{AB} X_i + G_Y^{A'B} Y_i + q_{XY}^{AB} \cdot X_i \cdot Y_i \qquad (1.21)$$

It is evident then that the four-parameter LFER, equation 1.9, must be an approximation. If it is observed to hold with high precision, a further condition, (iii) $(\partial^2 \Delta F^B/\partial x \partial y) = 0$, is required. There are, however, two terms in equation 1.9 on which deviations may be "smoothed out", and, if the range of variation of either x or y is not large, the cross-product, the third term on the right-hand side of equation 1.21, may not cause an obvious failure of the four-parameter correlation.

Similar considerations apply to modifications of the two-parameter LFER such as

$$\log(k_i^B/k_0^B) = G_X^{AB} \cdot [X_i + (\partial X_i/\partial y)(y_B - y_A)] \tag{1.22}$$

which seeks to account deviations from equation 1.8 due to additional variations in reaction series B or

$$\log(k_i^B/k_0^B) = X_i[G_{X_0}^{AB} + (\partial G_X^{AB}/\partial x)(x_i - x_0)] \tag{1.23}$$

which seeks to account for variations in $(\partial \Delta F^A/\partial x)$, $(\partial \Delta F^B/\partial x)$ or both with x in the x_0 to x_i range.

Multi-parameter correlations require a more critical assessment than two-parameter correlations. Additional parameters inevitably improve the correlations without necessarily providing additional information. Further, this information is often difficult to extract, and may be ambiguous. The parameters obtained are only useful as theoretical tools when the variables themselves have been identified and independent evidence clearly indicates that they are involved.

REFERENCES

Brønsted, J. N., and Pederson, K. J. (1924). Z. phys. Chem., **108**, 185.
Dippy, J. F. J., and Watson, H. B. (1936). J. chem. Soc., 436.
Grunwald, E., and Winstein, S. (1948). J. Am. chem. Soc., **70**, 846.
Hammett, L. P. (1937). J. Am. chem. Soc., **59**, 96.
Hammett, L. P., and Pfluger, H. L. (1933). J. Am. chem. Soc., **55**, 4079.
Hixon, R. M., and Johns, J. B. (1927). J. Am. chem. Soc., **49**, 1786.
Kindler, K. (1928). Justus Liebigs Annln Chem., **464**, 278.
Nathan, W. S., and Watson, H. B. (1933). J. chem. Soc., 890.
Ostwald, W. (1885). J. prakt. Chem., **31**, 433.

2

Correlation of Structural Changes in the Substrate

It is convenient to view a reaction in terms of three parts—the substrate, the reagent (or reagents), and the reaction medium. The designation of one reactant as substrate and the remainder as reagents is arbitrary, although the more complex reactant is normally regarded as the substrate. Clearly any reactant not consumed in the overall reaction, i.e. a catalyst, will be termed a reagent.

Structural variations in the substrate may be made in a relatively subtle manner in parts of the molecule remote from the site of reaction. Typical examples are the change from benzoic acid to *m*-nitrobenzoic acid, or from cyclohexylamine to 4-chlorocyclohexylamine. A more drastic change would be from acetic acid to chloroacetic acid, or from methylamine to aniline.

The Hammett equation is concerned with the influence of substituents on the reactivity of aromatic substrates. In its original form it is appropriate only to *m*- and *p*-substituted compounds where the reaction site is separated from the aromatic group by a non-conjugating side chain. There have been several extensions and modifications beyond these limitations.

The Taft equation performs a similar function with respect to non-aromatic substrates. It is more limited in the number and type of reaction series correlated, but in some respects the structural changes are more drastic.

2.1 THE HAMMETT EQUATION

Hammett first suggested in 1937 that an equation of the form

$$\log(k/k_0) = \sigma\rho \qquad (2.1)$$

might be employed to correlate the influence of *m*- and *p*-substituents on the reactivity of substrates containing aromatic groups. Thirty-nine reaction series and thirty substituents were examined in the first paper. Subsequently this was extended to fifty-two reaction series when a detailed discussion of the application of equation 2.1 was presented in 1940.

The parameter, σ, in equation 2.1 is *defined* by

$$\sigma = \log K_a - \log K_a^0$$
$$= pK_a^0 - pK_a = \Delta pK_a \qquad (2.2)$$

where K_a and K_a^0 are the dissociation constants for a substituted benzoic acid and benzoic acid itself, respectively, measured in water solution at 25°C.

When a linear relationship is observed between the corresponding σ values and the logarithms of some rate or equilibrium constants, i.e. equation 2.1 holds, then ρ is simply the slope of the line obtained. An example is illustrated in Fig. 2.1.

Clearly the substituent parameter, σ, depends upon the substituent and is independent of the reaction series correlated if equation 2.1 holds. Similarly, the reaction parameter, ρ, depends upon the reaction series, but not upon the substituents employed.

The quantity correlated in equation 2.1 is the *relative* reactivity of the substrate. The influence of the substituent is compared firstly with that of the standard substituent, H, and secondly with its relative effect on the standard reaction, i.e. that used to define the σ values. Absolute reactivities are obtainable from equation 2.1 provided that σ, ρ and $\log k_0$ are known, and from the equivalent equation 2.3

$$\log k_X - \log k_Y = (\sigma_X - \sigma_Y)\rho \qquad (2.3)$$

provided σ_X, σ_Y, ρ and $\log k_Y$ are known.

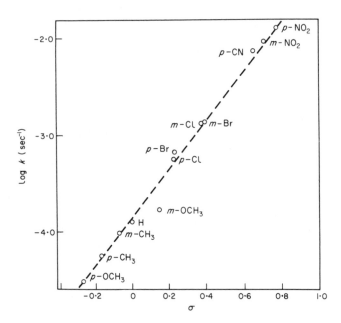

FIG. 2.1. Methanolysis of *m*- and *p*-substituted (–)-menthyl benzoates at 40°C: slope, $\rho = +2\cdot55$; intercept, $\log k_0 = -3\cdot86$.

It is important to recognize that at this stage nothing has been said about the changes in the "true" variable as opposed to the observed variations, i.e. ΔpK_a or $\log(k/k_0)$. As discussed in Chapter 1, the existence of the LFER, equation 2.1, implies that either a single variable is involved or, if there are changes in several variables, then these changes are proportional to one another in a manner independent of the particular substrate and reaction series. Further, the σ values, i.e. ΔpK_a, provide measures of these changes and are quantitatively useful in several reaction series.

Three observations suggest the nature of the "true" variables. Firstly, equation 2.1 only applies to m- and p-substituents. It fails in the case of o-substituents and, in most cases, similar reactions of substituted aliphatic substrates cannot be correlated with the relative strengths of the substituted acetic acids. Secondly, equation 2.1 fails for certain p-substituted aromatic substrates for which there are reasons to believe that during reaction substantial changes take place in direct conjugative interactions between the substituent and the reaction site. Thirdly, the reaction series that can be correlated are of the polar type and the magnitude of the ρ value is larger the closer the reaction site is to the aromatic group. Thus in order that equation 2.1 shall hold the substituent changes must be made at positions remote from the reaction site, where conjugative interactions do not take place and the reaction must be sensitive to changes in polar character.

A substituent may interact with a reaction site either directly by *proximity effects*, e.g. hydrogen bonding, neighbouring-group involvement, steric retardation or steric acceleration, by *conjugative interactions*, i.e. resonance effects arising from delocalizations involving both the substituent and the reaction site, by *direct electrostatic interactions*, arising from substituent poles or dipoles, or indirectly by means of *polar effects*. The latter are envisaged to arise from the distortion of the σ bonded and π bonded electrons of the substrate (I_σ and I_π *inductive effects*) due to the relative electronegativity of the substituent, and from distortions of the π electron distribution due to conjugative interactions between them and suitable orbitals on the substituent (*resonance polar effect*).

The observed restrictions on equation 2.1 suggest that the only permissible variations are in the direct electrostatic interactions and the polar effects.

Many successful correlations have been achieved by using equation 2.1. In 1953 Jaffé was able to list 204 reactions, many under several temperature and solvent conditions, yielding more than 350 reaction series. A critical re-examination was carried out with the σ values given by Hammett, not all of which arose through equation 2.2. Correlations were carried out by statistical methods and full details of ρ values, correlation coefficients (r) and other information concerned with the precision of the Hammett equation were reported. Using the ρ values of well correlated series, ($r > 0.95$), further

σ values were derived as those best fitting the entire body of experimental data. Marked deviations from equation 2.1 were in some cases obvious and could be accommodated by enhanced σ values as suggested by Hammett. Smaller deviations however tended to be "smoothed out" by the statistical procedure.

Acid-strengthening p-substituents, e.g. NO_2, CN, CO_2R, which are capable of conjugative interactions ($+ R$ groups) had been found by Hammett to require a special σ value for reactions of phenols, anilines and their derivatives. For example, the p-NO_2 group requires $\sigma = 1\cdot27$ whereas the defined value is $0\cdot78$. A corresponding duality of σ values, though expected, was not observed for conjugative acid weakening p-substituents ($- R$ groups), e.g. NR_2, OR. In this case a marked variability with reaction type was noted.

It is apparent, however, that the standard reaction series and many of the supposedly side-chain reactions involve reacting groups of the CO_2R type. For these direct conjugative interaction between the reaction site and donor substituents in the p-position is possible. This being so, σ values for these groups defined by equation 2.2 will in fact be enhanced values. With this in mind, van Bekkum et al. (1959) suggested that a limited set of "primary" σ values be defined by equation 2.2 and that these be used to establish correlations by equation 2.1. The set comprises only m-substituents, and excludes substituents of the type NR_2 and OR, where some indirect conjugative enhancement may be possible. To extend the set, two p-substituents are also included to be used only when direct conjugation is considered not possible.

At about the same time Taft (1960) had reached similar conclusions and suggested that correlations be based upon a select set of m-substituents. By choosing only reaction series where at least one methylene group "insulates" the reaction site from the aromatic system, e.g. reaction series 4,

TABLE 2.1. "PRIMARY" σ VALUES

ΔpK_a; Substituted benzoic acids: Water; 25°C					
Substituent	σ	σ^0	Substituent	σ	σ^0
1. m-CH_3	$-0\cdot07$	$-0\cdot07$	7. m-$COCH_3$	$0\cdot38$	$0\cdot34$
2. H	$0\cdot00$	$0\cdot00$	8. m-NO_2	$0\cdot71$	$0\cdot70$
3. m-F	$0\cdot34$	$0\cdot35$	9. p-$COCH_3$	$0\cdot50$	$0\cdot46$†
4. m-Cl	$0\cdot37$	$0\cdot37$	..	..	$0\cdot40$‡
5. m-Br	$0\cdot39$	$0\cdot38$	10. p-NO_2	$0\cdot78$	$0\cdot82$†
6. m-I	$0\cdot35$	$0\cdot35$	..	..	$0\cdot73$‡

† For reaction series in water or aqueous solvent mixtures. ‡ For reaction series in non-hydroxylic solvents.

TABLE 2.2. REACTION PARAMETERS AND CORRELATION DATA

Reaction series	n^a	ρ	r^b	$-(\log k_0)^c$
Acid dissociations				
1. $ArCO_2H$; H_2O; 25°	..	1·00	..	..
2. $ArCO_2H$; 50% EtOH; 25°	5	1·57±0·07	0·997	5·78
3. $ArCO_2H$; benzene; 25°	5	2·15†	..	..
4. $ArCH_2CO_2H$; H_2O; 25°	5	0·56±0·16§	0·982	4·34
5. $ArCH_2CH_2CO_2H$; H_2O; 25°	4	0·24±0·01	0·997	3·66
6. $ArCH{=}CHCO_2H$; H_2O; 25°	4	0·42±0·03	0·995	4·43
7. $ArC{\equiv}CCO_2H$; 50% EtOH; 24°	4	0·63±0·00	1·000	3·40
8. $ArB(OH)_2$; 25% EtOH; 25°	7	2·18±0·00	1·000	9·70
9. $ArPO_3H_2$; H_2O; 25°	5	0·75±0·00	1·000	1·83
10. $ArPO_3H^-$; H_2O; 25°	5	1·11±0·04	0·998	7·08
11. $ArOH$; H_2O; H_2O; 25°	7	2·26±0·07	0·997	9·95
12. $ArOH$; 49% EtOH; 21°	6	2·69±0·24	0·984	11.26
13. $ArSH$; 49% EtOH; 21°	6	2·45±0·13	0·994	7·80
14. $ArNH_3$; H_2O; 25°	7	2·94±0·06	0·999	4·66
15. $ArNH_3$; 30% EtOH; 25°	5	3·19±0·17	0·995	4·11
16. $ArN(CH_3)_2H^+$; H_2O; 20°	3	3·56±0·00	1·000	5·18
17. $ArCH_2NH_3^+$; H_2O; 25°	7	1·05‡	0·997	9·39
18. $ArNH.NH_3^+$; H_2O; 25°	5	1·17‡	..	5·20
Reaction between				
19. $ArCO_2H + CH_3OH + H^+$; 25°	7	−0·58±0·04	0·988	3·70
20. $ArCO_2H + Ph_2CN_2$; 60% acetone; 30°	5	0·88±0·00	1·000	1·80
21. $ArCO_2CH_3 + OH^-$; 60% acetone; 25°	5	2·38±0·19	0·991	2·14
22. $ArCO_2C_2H_5 + OH^-$; 60% acetone; 25°	7	2·47±0·10	0·996	2·62

5, 24 and 30 of Table 2.2, a set of σ^0 values has been proposed. No pathway for direct conjugative interactions is available in these substrates.

Table 2.1 lists the suggested "primary" σ values and gives for comparison corresponding σ^0 values. A very close correspondence between the two sets is observed bearing in mind the suggested precision of the σ^0 values, ±0·03. Despite the fact that the ΔpK_a values, which define the primary σ values, are certainly known to a precision better than ±0·01, this precision does not extend to any derived σ values, (see Table 2.3), and there is little to choose between the σ and σ^0 values on this score. In setting up the σ^0 scale some solvent dependence was observed as indicated in Table 2.1. The σ values are, of course, defined by a reaction series carried out in aqueous solution.

Some of the correlations achieved by means of the primary σ values are listed in Table 2.2. With this substituent restriction, only 60 series of Jaffé's listing, i.e. about 16%, and a further twenty series were examined. Of reaction series common to the two listings, it is found that the primary σ values yield 98% having $r > 0·98$ and 34% having $r > 0·998$, whereas Jaffé found only

Table 2.2. Reaction Parameters and Correlation Data (Continued)

Reaction series	n^a	ρ^b	r^b	$-(\log k_0)^c$
23. $ArCO_2C_2H_5 + OH^-$; 85% EtOH; 25°	5	$2\cdot61 \pm 0\cdot08$	0·999	3·26
24. $ArCH_2CO_2C_2H_5 + OH^-$; 60% acetone; 25°	8	1·00	..	1·36
25. $ArCH=CHCO_2C_2H_5 + OH^-$; 85% EtOH; 25°	5	$1\cdot24 \pm 0\cdot00$	1·000	2·86
26. $ArCOCl + CH_3OH$; 0°	6	$1\cdot42 \pm 0\cdot09$	0·992	3·31
27. $ArCOCl + C_2H_5OH$; 0°	7	$1\cdot74 \pm 0\cdot07$	0·996	4·11
28. $ArCONH_2 + OH^-$; 60% EtOH; 53°	4	$1\cdot40 \pm 0\cdot06$	0·998	5·12
29. $ArCONH_2 + H^+$; 60% EtOH; 52°	4	$-0\cdot50 \pm 0\cdot03$	0·996	5·60
30. $ArCH_2OCOCH_3 + OH^-$; 60% acetone; 25°	4	$0\cdot75 \pm 0\cdot05$	0·996	1·18
31. $ArCH_2OTos + H_2O$; 50% acetone; 25°	7	$-2\cdot32 \pm 0\cdot25$	0·972	3·58
32. $ArCH_2Cl + H_2O$; 50% acetone; 60°	7	$-1\cdot31 \pm 0\cdot16$	0·964	5·21
33. $Ar_2CHCl + EtOH$; 25°	5	$-4\cdot03 \pm 0\cdot22$	0·996	4·25
34. $ArCMe_2Cl + H_2O$; 90% acetone; 25°	8	$-4\cdot48 \pm 0\cdot12$	0·998	3·95
35. $ArNH_2 + HCO_2H$; 67% pyridine; 100°	5	$-1\cdot30 \pm 0\cdot06$	0·997	$-0\cdot58$
36. $ArNH_2 + PhCOCl$; C_6H_6; 25°	5	$-3\cdot21 \pm 0\cdot09$	0·999	1·12
37. $ArN(CH_3)^2 + CH_3I$; acetone; 35°	4	$-2\cdot39 \pm 0\cdot00$	1·000	3·69
38. $ArOCOCH_3 + OH^-$; 60% acetone; 15°	3	$1\cdot51 \pm 0\cdot00$	1·000	0·62
39. $ArOSiEt_3 + OH^-$; 51% EtOH; 25°	3	$1\cdot99 \pm 0\cdot09$	0·999	$-0\cdot15$
40. $ArOSiEt_3 + H^+$; 51% EtOH; 25°	3	$-0\cdot47 \pm 0\cdot03$	0·998	0·65
41. $ArH + NO_2^+$; Ac_2O; 25°	5	$-7\cdot29 \pm 0\cdot65†$	0·988	..
42. $ArB(OH)_2 + Br_2$; 20% HOAc; 25°	7	$-3\cdot83 \pm 0\cdot21$	0·992	2·23
43. $Ar_3C^+ + H_2O \rightleftarrows Ar_3COH$; 25°	6	$3\cdot97 \pm 0\cdot15$	0·997	6·76

† Relative log k values correlated.
‡ Correlation with σ^0.
§ Exclusion of p-NO_2 datum yields an improved correlation with $\rho = 0\cdot50$.

a. Number of log k values correlated.
b. Correlation coefficient.
c. Intercept of the regression line.

59% with $r > 0\cdot98$ and none having $r > 0\cdot998$. Indeed in only 6% of all the series correlated by Jaffé was this high precision observed. The improvement clearly arises from the exclusion of "abnormal" effects arising in the case of some p-substituents.

Employing the ρ values of well correlated reaction series, $r > 0\cdot98$, "normal" substituent parameters, σ^n, have been computed by excluding those reaction series for which direct conjugative effects may be involved. An essentially equivalent procedure is the derivation of σ^0 values using ρ values based upon the select group of m-substituents.

The σ^n values are listed in Table 2.3 and the σ^0 values in Table 2.4. Again the two sets are very similar except in the case of m- and p-$N(CH_3)_2$, m- and p-NH_2 and p-F. It can be argued that most of the reaction series employed to derive the σ^0 values have small ρ values, and hence low sensitivity to

TABLE 2.3. "NORMAL" SUBSTITUENT PARAMETERS (σ^n)

Substituent	σ_m	σ_p
OH	$0\cdot10\pm0\cdot03$	$-0\cdot18\pm0\cdot04\dagger$
OCH_3	$0\cdot08\pm0\cdot04$	$-0\cdot11\pm0\cdot02\dagger$
NH_2	$-0\cdot04\pm0\cdot04$	$-0\cdot17\pm0\cdot03\dagger$
$N(CH_3)_2$	$-0\cdot05\pm0\cdot09$	$-0\cdot17\pm0\cdot08\dagger$
$N(CH_3)_3^+$	$0\cdot86\pm0\cdot09$	$0\cdot80\pm0\cdot09$
CH_3	$..$	$-0\cdot13\pm0\cdot04$
C_2H_5	$(-0\cdot07)\ddagger$	$-0\cdot12\pm0\cdot03$
$i\text{-}C_3H_7$	$(-0\cdot07)\ddagger$	$-0\cdot10\pm0\cdot03$
$t\text{-}C_4H_9$	$(-0\cdot07)\ddagger$	$-0\cdot14\pm0\cdot03$
CO_2H	$0\cdot35\pm0\cdot02$	$0\cdot40\pm0\cdot04$
CO_2^-	$0\cdot07\pm0\cdot11$	$0\cdot12\pm0\cdot04$
CO_2CH_3	$0\cdot32\pm0\cdot04$	$0\cdot46\pm0\cdot02$
CN	$0\cdot61\pm0\cdot04$	$0\cdot67\pm0\cdot04$
CF_3	$0\cdot47\pm0\cdot07$	$0\cdot53\pm0\cdot08$
$Si(CH_3)_3$	$-0\cdot05\pm0\cdot06$	$0\cdot01\pm0\cdot03$
SCH_3	$0\cdot23\pm0\cdot02$	$(0\cdot22)\ddagger$
$S(CH_3)_2^+$	$(1\cdot03)\ddagger$	$(1\cdot20)\ddagger$
SO_2CH_3	$0\cdot68\pm0\cdot04$	$(0\cdot67)\ddagger$
F	$..$	$0\cdot06\pm0\cdot03$
Cl	$..$	$0\cdot24\pm0\cdot03$
Br	$..$	$0\cdot27\pm0\cdot03$
I	$..$	$0\cdot30\pm0\cdot03$

† Reaction series involving anilines and aniline derivatives
  also excluded.
‡ From reaction series some of which have $r < 0\cdot98$.

substituent effects. This means that the σ^n values are probably more precise, and one is more certain of their wide range of application. On the other hand the σ^0 values are certainly free from direct resonance contributions, so that the observed differences noted above for donor groups probably arise from overcompensation of the σ^n values. This has presumably arisen since, although aniline reactions were omitted, phenol and thiophenol reactions were employed to derive σ^n.

With the exception of the charged substituents and the $N(CH_3)_2$ group the σ^n values for most m- and p-substituents listed in Table 2.3 have a precision of about $\pm0\cdot04$. These exceptions may arise from failure to eliminate direct conjugative interactions completely, although variable solvent effects may be the most serious cause of deviations. Within this precision, all alkyl groups have $\sigma_m = -0\cdot07$ and $\sigma_p = -0\cdot12$. Similarly all the halogens have $\sigma_m = 0\cdot36$ and all except F have $\sigma_p = 0\cdot27$.

Solvent variations were considered in the derivation of the σ^0 values and the solvent effect is indicated in Table 2.4. Of all the reaction series correlated

TABLE 2.4. THE σ^0 PARAMETERS (ESTIMATED PRECISION ± 0.03)

Substituent	σ_m^0	σ_p^0
$N(CH_3)_2$	-0.15	-0.44
NH_2	-0.14	-0.38
OCH_3	$\begin{cases} 0.13^a \\ 0.06^b \end{cases}$	$\begin{cases} -0.12^a \\ -0.16^b \end{cases}$
OH†	0.04^c	-0.13^c
SCH_3	0.10	0.02
CH_3	-0.07	-0.15
F	0.35	0.17
Cl	0.37	0.27
Br	0.38	0.26
I	0.35	0.27
CO_2R	0.36	0.46^d
$COCH_3$	0.34	$\begin{cases} 0.46^d \\ 0.40^c \end{cases}$
CF_3	0.48	0.54
CN	0.62	$\begin{cases} 0.69^d \\ 0.63^c \end{cases}$
NO_2	0.70	$\begin{cases} 0.82^d \\ 0.73^c \end{cases}$

† σ is very strongly solvent dependent for aqueous solvent mixtures.
a. For aqueous solutions only.
b. For non-hydroxylic media and most aqueous solvent mixtures.
c. For non-hydroxylic solvents.
d. For water and aqueous solvent mixtures.

by the primary σ values, 46% were carried out in a solvent that is at least half water and in another 40% of the series the solvent is an alcohol or a partially aqueous medium. Solvent effects are thus not clearly apparent and the σ^n values are aqueous solvent values.

Table 2.5 lists some further σ values obtained from various sources which must be regarded as approximate indications of the effect of these substituents.

Having concluded that the measured relative substituent effect arises only from direct electrostatic interactions and polar effects, the normal substituent parameters, σ^n or σ^0, can be employed as quantitative measures of these effects. Because of the definition of σ (equation 2.2), the magnitude and sign of the σ value is simply the relative substituent effect on benzoic acid dissociation in water at 25°C. Any theory of σ values must be a theory of substituent effects on acid dissociation. However the wide range of application of

TABLE 2.5. OTHER SUBSTITUENT PARAMETERS

Substituent	σ_m	σ_p	Substituent	σ_m	σ_p
C_6H_5	0·06	−0·01	AsO_3H^-	..	−0·02
CCl_3	0·40	0·46	OC_6H_5	0·25	−0·32
CH_2CN	..	0·01	$OCOCH_3$	0·39	0·31
CH_2Cl	..	0·18	OCF_3	0·4	0·5
CHO	0·35	0·22	O^-	−0·2	−0·5
$CONH_2$	0·28	..	SH	0·25	0·15
$C{\equiv}CC_6H_5$	0·14	0·16	$S.COCH_3$	0·39	0·44
$CH_2Si(CH_3)_3$	−0·16	−0·21	SCN	..	0·52
$Si(C_2H_5)_3$	..	0·00	$SOCH_3$	0·52	0·49
$Ge(C_2H_5)_3$	..	0·00	SO_2NH_2	0·46	0·57
$Sn(C_2H_5)_3$	..	0·00	SO_3H	..	0·50
$NHNH_2$	−0·1	−0·4	SO_3^-	0·05	0·09
$NH.COCH_3$	0·25	0·04	SCF_3	0·44	0·57
N_3	0·33	0·08	SCF_5	0·6	0·7
$N(CF_3)_2$	0·45	0·53	$SeCH_3$	0·2	0·0
$P(CH_3)_2$	0·1	0·05	IO_2	0·70	0·76
$P(CF_3)_2$	0·6	0·7	$B(OH)_2$	0·01	0·45
$P(CH_3)_3^+$	0·8	0·9			
PO_3H^-	0·2	0·26			

the σ values to other equilibria and to rate processes means that such a theory of acid dissociation can become a theory of substituent effects in general.

To be able to make any progress in the use of empirical correlations in the development of a theory of substituent effects, it is necessary to assume that to a good approximation the total substituent effect can be treated as the sum of contributions from the various factors discussed above. Any other approach would be mathematically and conceptually too difficult.

Qualitatively the σ values conform to the general picture of substituent effects with electron-withdrawing groups being acid strengthening, σ is positive, and electron releasing groups acid weakening, σ is negative. More quantitatively the difference $\sigma_p - \sigma_m$ can be taken as a measure of conjugative interactions between the substituent and the aromatic system. Groups such as NO_2, CN and CO_2R appear to exert a small electron-withdrawing effect by this mechanism, and larger electron-releasing effects are observed for donor groups in the order $NR_2 >$ OR $>$ F $>$ Cl, Br $>$ I, CH_3. For the halogens, $\sigma_m > \sigma_p$ is in agreement with an electron-withdrawing non-conjugative effect $(+I)$ exceeding an electron-releasing conjugative effect $(-R)$, and for p-NR_2, m-NR_2 and p-OR groups the order $-R > +I$ agrees with the general qualitative theory of substituent effects.

In order to achieve a quantitative separation of the total substituent effect Taft and Lewis (1959) have suggested the following

$$\sigma_m = \sigma_I + \alpha\sigma_R \qquad (2.4)$$

$$\sigma_p = \sigma_I + \sigma_R \qquad (2.5)$$

where σ_I measures the contributions to the σ value arising from non-conjugative (*inductive*) and σ_R the contribution from conjugative (*resonance*) interactions between the substituent and the aromatic system. The factor α allows for the lower effectiveness of conjugative interactions from *m*- as compared to *p*-dispositions.

This type of treatment is considered in more detail in Chapter 3, so that for the present it suffices to point out the more important assumptions and approximations involved. These are

(a) that the total substituent effect can be expressed as the sum of inductive and resonance contributions;
(b) that the inductive contributions to σ_m and σ_p are equal;
(c) that a unique proportionality factor, α, can be found; and
(d) that σ_I values are obtainable from some other source.

If σ_I and σ_R depend only upon substituent and the same ρ value is applicable to *m*- and *p*-reaction series (see below) then

$$\sigma_I = (\sigma_m - \alpha\sigma_p)/(1 - \alpha) \qquad (2.6)$$

and

$$(1 - \alpha)^{-1}[\log(k^m/k_0) - \alpha \log(k^p/k_0)] = \sigma_I\rho \qquad (2.7)$$

From correlations of polar effects in aliphatic substrates by the Taft equation (see below) a set of substituent parameters has been obtained that appear to be equal to the σ_I values required in equations 2.4–2.7. It is found that equation 2.7 is followed not only by reaction series correlated by the σ^n or σ^0 values but also those for which *p*-substituent effects are "abnormal". Initially two values of α were employed, $\frac{1}{3}$ for "normal" reaction series and $\frac{1}{10}$ for "abnormal" series. However, Roberts and Jaffé (1959) have shown that highly significant improvements in correlation result from permitting α to be an adjustable parameter. This is equivalent to using two reaction parameters ρ_I and ρ_R appropriate to the σ_I and σ_R dependencies of $\log(k/k_0)$, respectively. The usual reaction parameter is then ρ_I and $\alpha = \rho_R/\rho_I$.

For each aromatic reaction series there are only two observable quantities, namely, the *m*- and *p*-substituent effects and hence the maximum number of unknowns that can be handled is two. For a fixed value of α these are σ_I and σ_R. However if a further observable, the aliphatic substituent effect, is employed to specify σ_I, then α may also be variable.

The fact that equation 2.7 gives good correlations where equation 2.1 is sometimes unsuccessful suggests the basic assumptions of the treatment are at least good approximations. Deficiencies, such as unequal m- and p-inductive contributions or the neglect of other factors, will become included in some way in the statistically derived values of σ_R and α. Neither of these is found to be constant and, although in both cases this is not unexpected, the variations in α are not always readily explicable. Nevertheless correlations of physical data, particularly from n.m.r. spectroscopy (see Chapter 6), have tended to substantiate the σ_I and σ_R values as useful quantitative measures of factors contributing to the total substituent effect and support the assumptions as to their origin.

Taft's method of separation is a purely empirical procedure although guided by an assumption as to the nature of the two independent and additive factors involved. Dewar and Grisdale (1962) have suggested another two-factor separation based upon the identification of these factors as direct electrostatic (*field*) effects and π-electron (*mesomeric*) effects. The unknowns in this case are the parameters F and M, constants for a given substituent, which determine the *field* and *mesomeric* effects, respectively. The field effect is considered to depend upon $1/r_{ij}$, where r_{ij} is the distance between the points of attachment to the aromatic system of the substituent and the side chain. The mesomeric effect, which includes the π inductive as well as the resonance polar effect, is made dependent upon π_{ij}, the atom–atom polarizability of the aromatic system. Thus

$$\sigma = F/r_{ij} - M\pi_{ij} \qquad (2.8)$$

In this method a rather precise model for the mode of operation of substituent effects is set up and using the σ_m and σ_p values the required values of F and M are derived. When these were used and equation 2.8 employed to calculate σ values for other aromatic systems considerable success is obtained but with notable exceptions. As described below, the results for certain positions in the naphthalene system indicate serious deficiencies in the $1/r_{ij}$ term and the neglect of other effects.

There is a rough correspondence between the F values and σ_I and between the M values and σ_R (see Table 2.6). However the strong donor groups, F, OH, OCH_3 and NH_2, appear to have been assigned F values ca. 0·5 units too small. In general plots of F versus σ_I and M versus σ_R display the considerable scatter that must arise from the two different ways of analysing the total substituent effect. It may be that the σ_I values express both field effects and π-inductive effects whereas these are separated into respectively the F and M terms through equation 2.8. In this case a non-conjugating substituent (σ_R small) could still influence the π electron system (M not small). If the field effect and the π inductive effects were roughly proportional, a usefully

TABLE 2.6. COMPARISON OF TAFT AND DEWAR–GRISDALE PARAMETERS

Substituent	σ_I	F	σ_R	$-M$
$(CH_3)_3N^+$	0·90	1·53	−0·02	0·51
NO_2	0·63	1·26	0·15	1·48
CH_3SO_2	0·62	1·08	0·10	1·83
CH_3SO	0·56	0·97	−0·07	0·37
CN	0·56	1·00	0·10	1·60
F	0·52	0·56	−0·46	−2·10
Cl	0·47	0·63	−0·24	−0·91
Br	0·45	0·66	−0·22	−1·00
CF_3	0·42	0·77	0·12	1·46
I	0·40	0·59	−0·22	−1·10
CO_2R	0·30	0·66	0·15	1·19
HO	0·30	0·12	−0·67	−4·38
CH_3O	0·30	0·13	−0·57	−3·36
CH_3CO	0·28	0·68	0·22	1·62
HS	0·23	0·42	−0·08	−0·64
CH_3S	0·19	0·24	−0·19	−1·19
NH_2	0·13	−0·38	−0·79	−4·75
CH_3	−0·05	−0·14	−0·12	−1·00
$(CH_3)_3Si$	−0·11	−0·08	0·04	−0·33

constant scale of σ_I values could be set up that would parallel the F values. In addition, any deficiencies in the model, i.e. in the $1/r_{ij}$ and π_{ij} dependencies, will be smoothed out in the statistical treatment by adjustments in F and M.

Whatever the precise mode of operation of substituent effects, the σ^n or σ^0 values represent quantitative measures of "normal" effects. "Abnormality" can be assessed in terms of the difference between the observed and calculated values of $\log(k/k_0)$, for example by ψ in equation 2.9

$$\log(k/k_0) - \sigma^0\rho = \psi \neq 0 \qquad (2.9)$$

When calculated values of σ, $= \log(k/k_0)/\rho$, are derived for a large number of reaction series, most m-substituents have values close to σ^n, but a wide range of values are obtained for most p-substituents. There is an essentially continuous population of this range for each substituent with little indication of any grouping about an enhanced value. For certain groups of reaction series, the special parameters have some utility, e.g. σ^+ for electrophilic nuclear substitutions, and σ^- for nucleophilic substitutions, aniline and phenol reactions. Overall it is clear that different reactions lead to different degrees of enhancement.

From the nature of the substituents involved and the reactions taking place, these enhancements can often be identified as arising from changes in direct conjugative interactions. Equation 2.10 has been suggested as the means of estimating these changes

$$\Delta\Delta F_R = -2 \cdot 303 R T (\sigma - \sigma^n)\rho \tag{2.10}$$

The σ^0 values may be used in essentially the same way, and if it is believed that only resonance contributions are involved in the "abnormal" substituent effect

$$\sigma - \sigma^0 = \sigma_R - \sigma_R^0 = \Delta\sigma_R \tag{2.11}$$

may also be employed as a measure of the abnormality.

In all these equations, 2.9–2.11, it is necessary to assume that the same ρ value applies to m- and p-series. As discussed below, this may not always be valid.

The meaning attached to the magnitude and sign of the reaction parameter, ρ, depends upon the interpretation given to the σ values. The LFER simply yields ρ as a proportionality factor that depends upon the response of a particular reaction series to substituent changes. Clearly, from the definition of σ, reactions facilitated by acid-strengthening groups will have positive ρ values, and reactions more sensitive to substituent effects than the standard reaction will have ρ values numerically greater than unity.

Having associated electron withdrawal with acid-strengthening behaviour, it follows that a positive ρ value implies the development of negative charge (or loss of positive charge) at the reaction site. The larger the change in polarity at the reaction site and the closer the reaction site is to the aromatic system, the larger will be the magnitude of the ρ value.

In terms of the generalized LFER discussed in Chapter 1, the definition of the substituent parameters is

$$\sigma_i = -(\Delta F_i^A - \Delta F_0^A)/2 \cdot 303 R T_A \tag{2.12}$$

where ΔF^A is the free-energy change in the dissociation of a benzoic acid in water at $25°C(T_A)$. Hence the reaction parameter for some reaction series B, following equation 1.14, is

$$\rho_B = G_\sigma^{AB} = \frac{T_A}{T_B} \frac{(\partial\Delta F^B/\partial\sigma)}{(\partial\Delta F^A/\partial\sigma)}$$

$$= \frac{G_0}{T_B} \cdot (\partial\Delta F^B/\partial\sigma) \tag{2.13}$$

where G_0 is a constant determined by the defining reaction series (A).

One immediate expectation based upon equation 2.13 is that if $(\partial\Delta F^B/\partial\sigma)$ is temperature invariant, then ρ should be inversely proportional to temperature, i.e.

$$\rho_1/\rho_2 = T_2/T_1 \tag{2.14}$$

In general $(\partial \Delta F^B / \partial \sigma)$ will not be temperature invariant unless the second term in equation 2.15 is zero, $(\partial \Delta S^B / \partial \sigma) = 0$, i.e. the reaction series is *isoentropic*.

$$\frac{\partial \Delta F^B}{\partial \sigma} = \frac{\partial \Delta H^B}{\partial \sigma} - T \frac{\partial \Delta S^B}{\partial \sigma} \qquad (2.15)$$

This is the case for a few reaction series, e.g. ester hydrolyses (reaction series 21 and 22 of Table 2.2). In these cases, equation 2.14 is followed with high precision.

There are many reaction series for which ρ increases essentially linearly with $1/T$ and yet $(\partial \Delta S^B / \partial \sigma)$ is not zero. In some of these cases *isokinetic* behaviour has been observed. This arises when the changes in ΔH and in ΔS appear to be linearly related as in equation 2.16

$$\Delta H^B = \Delta H_0^B + \beta \Delta S^B \qquad (2.16)$$

where β is the slope of the isokinetic line and ΔH_0^B is a substituent independent intercept.

It follows from equation 2.15 and 2.16 that

$$(\partial \Delta F^B / \partial \sigma) = (1 - T/\beta)(\partial \Delta H^B / \partial \sigma) \qquad (2.17)$$

and that

$$\rho = G_0 \left(\frac{1}{T_B} - \frac{1}{\beta} \right)(\partial \Delta H^B / \partial \sigma) \qquad (2.18)$$

Unfortunately isokinetic behaviour may be observed even when there is no *real* relationship between ΔH^B and ΔS^B. These quantities are usually calculated from the observed variation of equilibrium or rate constants with temperature, thus

$$\Delta H = \frac{-2 \cdot 303 R T_1 T_2}{(T_2 - T_1)}(\log k_2 - \log k_1)$$

and

$$\Delta S = \frac{2 \cdot 303 R}{(T_2 - T_1)}(T_2 \log k_2 - T_1 \log k_1)$$

If the temperature range is not large, e.g. $T_1 = 25°C$ and $T_2 = 55°C$, $T_2/T_1 = 328/298 \simeq 1 \cdot 1$, then

$$\frac{T_2}{T_1} \log k_2 - \log k_1 \simeq \log k_2 - \log k_1$$

and hence $\Delta H \simeq T_2 \Delta S \simeq T_1 \Delta S$. This will yield an isokinetic relationship with β having about the same magnitude as T_1 and T_2. This is indeed the commonly observed value of β.

In at least two reaction series, the substituent effect is manifested largely in ΔS changes with only small variations in ΔH, i.e. almost *isoenthalpic*. One of these is the standard reaction series.

It is often supposed that polar effects will influence only the ΔH term and leave the ΔS term unaffected. However, it is observed that very few of the reaction series correlated by the Hammett or Taft equations are isoentropic. Since the ΔS term is obtained from the temperature-dependent part of the free-energy change, then any polar effect that contains a temperature-dependent factor, such as dielectric constant, will make a contribution to the ΔS term.

Almost all the reaction series that have been correlated by the Hammett equation take place in solution. However, except for considerations of dielectric constants in direct electrostatic interactions and for allowances for variations in σ values, the solvent is not specifically taken into account in most discussions of substituent effects. Theoretical treatments deal with intrinsic effects of substituents and tend to become impossibly difficult when attempts are made to introduce solvent effects. On the other hand the empirical assessment of substituent effects by way of LFER does not indicate any serious error in this approximation. Hepler (1963) has suggested how this may arise.

The relative free-energy change that is employed to measure the substituent effect in some reaction is the free energy change in the exchange process

$$(\text{Reactants})_i + (\text{Products})_0 \rightleftharpoons (\text{Products})_i + (\text{Reactants})_0$$

and it can be considered composed of internal and external contributions as in equation 2.19.

$$\delta_R \Delta F = \Delta F_i - \Delta F_0$$
$$= \delta_R \Delta F_{\text{int}} + \delta_R \Delta F_{\text{ext}} \tag{2.19}$$

The internal interactions are those corresponding to intrinsic substituent effects and do not specifically involve solvent molecules, whereas the external factor arises from differences in solvent–solute interactions.

It is likely that for symmetrical reactions of this type the internal entropy change is zero, i.e. $\delta_R \Delta S_{\text{int}} = 0$, so that the resultant entropy change arises solely from external interactions, i.e. $\delta_R \Delta S = \delta_R \Delta S_{\text{ext}}$. Further, most models for solvent–solute interactions suggest that the entropy and enthalpy changes are approximately linearly related, e.g. by

$$\beta \delta_R \Delta S_{\text{ext}} \simeq \delta_R \Delta H_{\text{ext}} \tag{2.20}$$

It follows then that

$$\delta_R \Delta F \simeq \delta_R \Delta H_{int} + (1 - T/\beta)\delta_R \Delta H_{ext} \tag{2.21}$$

If β is approximately equal to T, as seems often to be the case, then $\delta_R \Delta F$, the observed substituent effect, will be approximately equal to $\delta_R \Delta H_{int}$, which is the quantity most useful from the theoretical standpoint.

It should be noted that equation 2.20 requires that isokinetic behaviour be observed (cf. equation 2.16), since

$$\beta \delta_R \Delta S = \beta \delta_R \Delta S_{ext} = \delta \Delta H_{ext}$$

$$= \delta_R \Delta H - \delta_R \Delta H_{int} \tag{2.22}$$

An alternative approach to the same problem can be based upon a hydrogen-bonding solvation model for aqueous solutions. It can be shown that for acid–base reactions $\delta_R \Delta F_a \simeq m \delta_R \Delta F_{gas}$ where m is temperature dependent and is determined by the relative strengths of hydrogen bonds between reactants and water and between products and water (cf. Chapter 4.3).

A few reaction series have been examined in a number of solvents. For acid dissociation, it is sometimes found that the relative acid strength is inversely proportional to the dielectric constant, D_s, of the solvent. Thus for substituted benzoic acids, equation 2.23 may be successful.

$$\Delta pK_s = \Delta pK_\infty + L/D_s \tag{2.23}$$

(ΔpK_∞ is the intercept corresponding to the relative acid strength in a solvent of infinite dielectric constant and L is a constant characteristic of the acid in question). Equation 2.23 is satisfactory when the solvent is water, methanol, ethanol or ethylene glycol. It is less successful, however, for propan-1-ol, butan-1-ol and for dioxan–water mixtures. In these cases the "bulk" dielectric constant appears to be inadequate. This conclusion is also reached when an attempt is made to account for substituent effects in terms of direct electrostatic interaction (see Chapter 3).

Equation 2.23 may be successful even when the bulk dielectric constant is inappropriate for the direct electrostatic factor. This can arise if the "effective" dielectric constant D_e, can be expressed in terms of D_s and the "internal" dielectric constant of the substrate, D_i, by

$$\frac{1}{D_e} = \frac{a}{D_s} + \frac{b}{D_i} + c \tag{2.24}$$

and the substituent effect by

$$\Delta pK = \frac{X}{D_e} + Y$$

$$= \frac{aX}{D_s} + \frac{bX}{D_i} + cX + Y \qquad (2.25)$$

since equations 2.23 and 2.25 are equivalent with

$$L = aX \qquad \text{and} \qquad \Delta pK_\infty = bX/D_i + cX + Y$$

Bearing this result in mind it is unlikely that any important significance can be attached to the value of ΔpK_∞.

In terms of the Hammett reaction parameter equation 2.23 may be re-expressed as equation 2.26

$$\rho_s = \rho_\infty + L'/D_s \qquad (2.26)$$

in which $L' = L/\sigma$ is now independent of the substrate. Equation 2.26 appears to be reasonably successful for the acidity of the benzoic acids and for the anilinium cations in water, methanol and ethanol, but is not successful when the changes involve the compositions of aqueous solvent mixtures. In these cases a plot of ρ_s versus solvent composition usually yields a smooth curve, from which new ρ_s values can be interpolated if required. More successful correlations of solvent variations are discussed in Chapter 4.

The magnitude of the reaction parameter is clearly associated with the position of the reaction site relative to the aromatic ring. For reactions involving a unit change in charge at an atom separated from the aromatic system by i "saturated" bonds $\rho \simeq (2.8 \pm 0.5)^{1-i}$. A more precise expression has been suggested by McGowan (1949) for acid dissociation at 25°C in water solution, $\rho = 2.5/2^i$ (Eq. I). The performance of this expression is illustrated in Table 2.7. The best expression of this type fitting the observed data is $\rho = 2.6/2.2^i$ (Eq. II). Although no simple expression can yield completely satisfactory predictions of reaction parameters, these equations are remarkably good. Their existence has important implications for any theory of the mode of transmission of substituent effects. More obviously they indicate how the magnitude of the ρ value may be employed in reaction mechanism and related studies. One example is reaction series 8 of Table 2.2. In this case $\rho = 2.18$ suggests that $i = 0$ and substantiates the view that the equilibrium is

$$ArB(OH)_2 + 2H_2O \rightleftharpoons ArB(OH)_3^- + H_3O^+$$

and not

$$ArB(OH)_2 + H_2O \rightleftharpoons ArBO_2H^- + H_3O^+$$

TABLE 2.7. CALCULATED REACTION PARAMETERS

Dissociating acids	i	ρ_{obs}	Eq. I	Eq. II
$XC_5H_4NH^+$	-1	6·11	5·0	5·7
$XC_6H_4NH_3^+$ XC_6H_4OH	0	2·94 2·26	2·5	2·6
$XC_6H_4CO_2H$ $XC_6H_4NH.NH_3^+$ $XC_6H_4CH_2NH_3^+$ $X.C_6H_4PO_3H_2$ $XC_6H_4PO_3H^-$	1	1·00 1·17 1·05 0·75 1·11	1·25	1·2
$XC_6H_4CH_2CO_2H$	2	0·56	0·625	0·54
$XC_6H_4CH_2CH_2CO_2H$	3	0·24	0·31	0·24

Since the magnitude and sign of the reaction parameter must depend largely upon the relative polar character of the initial and final (or transition) states of the reacting side chain, a dependence upon the σ values of these states would be expected. Hine (1959) has developed an expression for this dependence by considering the equilibria of equation 2.27.

$$X_1C_6H_4Y_1 \xrightleftharpoons{\kappa_1^Y} X_1C_6H_4Y_2$$
$$\kappa_1^X \Big\Updownarrow \qquad\qquad \Big\Updownarrow \kappa_2^X \qquad\qquad (2.27)$$
$$X_2C_6H_4Y_1 \xrightleftharpoons{\kappa_2^Y} X_2C_6H_4Y_2$$

If the unmodified Hammett equation can be applied to the effect of substituents X_1 and X_2 on the reaction represented by $ArY_1 \rightleftharpoons ArY_2$ (ρ_Y) and of Y_1 and Y_2 on $ArX_1 \rightleftharpoons ArX_2$ (ρ_X) then

$$\log(K_1^Y/K_2^Y) = (\sigma_{X_1} - \sigma_{X_2})\rho_Y$$
$$\log(K_1^X/K_2^X) = (\sigma_{Y_1} - \sigma_{Y_2})\rho_X$$

but since $K_1^Y/K_2^Y = K_1^X/K_2^X$ it follows that

$$\log K = (\sigma_{X_1} - \sigma_{X_2})\rho_Y = (\sigma_{Y_1} - \sigma_{Y_2})\rho_X \qquad (2.28)$$

where K is the equilibrium constant for

$$X_1C_6H_4Y_1 + X_2C_6H_4Y_2 \rightleftharpoons X_1C_6H_4Y_2 + X_2C_6H_4Y_1 \qquad (2.29)$$

Equation 2.28 is equivalent to

$$\rho_X/(\sigma_{X_1} - \sigma_{X_2}) = \rho_Y/(\sigma_{Y_1} - \sigma_{Y_2}) = \tau \qquad (2.30)$$

Where τ must be independent of the X and Y groups, it is determined by the reaction conditions, i.e. solvent and temperature, and by the relative disposition of X and Y in XC_6H_4Y. If the common assumption, that the same ρ value applies to both m- and p-derivations, is followed, then

$$\frac{\sigma_{m\text{-X}_1} - \sigma_{m\text{-X}_2}}{\rho_{p\text{-X}_1} - \sigma_{p\text{-X}_2}} = \frac{\tau_p}{\tau_m} = \frac{\sigma_{m\text{-CO}_2\text{H}} - \sigma_{m\text{-CO}_2^-}}{\sigma_{p\text{-CO}_2\text{H}} - \sigma_{p\text{-CO}_2^-}} \tag{2.31}$$

for all pairs of substituents that can be initial and final (or transition) states of reaction correlated by the unmodified Hammett equation. A survey of typical σ values reveals that equation 2.31 cannot hold and that τ_p/τ_m is not constant. However most substituent pairs, e.g. NO_2 and OCH_3 or F and CH_3, correspond either to unrealistic equilibria or to reactions involving nuclear substitutions. The unmodified Hammett equation is inapplicable in these cases. Further, the σ values corresponding to the pairs CO_2H/CO_2^-, OH/O^-, SH/S^-, and NH_3^+/NH_2 are the least certainly known of all substituents, and there are indications that the Hammett equation is not applicable to charged substituents.

In terms of individual substituents and reacting side chains Hine's suggestions are untestable. A statistical test has, however, been carried out. Jaffé (1959) finds that out of 336 reaction series the number for which ρ values correlating m- and p-derivatives are different is significantly greater than would be expected purely by chance. On the other hand in some 85% of the series the differences are not significant. McDaniel (1961) has pointed out how this situation can arise. Equation 2.31 is equivalent to a linear relationship between $\sigma_{p\text{-X}}$ and $\sigma_{m\text{-X}}$, and such relationships are observed for series of substituents having a common first atom, i.e. the one bonded to the aromatic system. Thus for the "carbon-family", CH_3, CO_2^-, CHO, CO_2H, CO_2R, $COCH_3$, CF_3, and CN and a variety of substituents of the type CH_2X, $\sigma_p \simeq 1{\cdot}15\sigma_m$. Thus provided that the reaction does not change the first atom of the group, equation 2.31 may be satisfied. It is important to note that this applies to the standard reaction for which the corresponding substituent pair is CO_2H/CO_2^-. It is also evident that $\tau_m/\tau_p \simeq 1{\cdot}15$.

A similar argument can be based upon the analysis of the substituent effect with inductive and resonance contributions.

$$\frac{\tau_m}{\tau_p} = \frac{(\sigma_I^p + \sigma_R)_{\text{X}_1} - (\sigma_I^p + \sigma_R)_{\text{X}_2}}{(\sigma_I^m + \alpha\sigma_R)_{\text{X}_1} - (\sigma_I^m + \alpha\sigma_R)_{\text{X}_2}}$$

$$= \frac{\Delta\sigma_I^p + \Delta\sigma_R}{\Delta\sigma_I^m + \alpha\Delta\sigma_R} \simeq \frac{\Delta\sigma_I^p}{\Delta\sigma_I^m} \tag{2.32}$$

if $\Delta\sigma_R = (\sigma_R)_{\text{X}_1} - (\sigma_R)_{\text{X}_2} \simeq 0$. This must be the case for a "true" side-chain reaction. If for any substituent $\sigma_I^p \simeq 1{\cdot}15\sigma_I^m$ then $\tau_m/\tau_p \simeq 1{\cdot}15$ as suggested above.

For any true side-chain reaction, $ArX_1 \rightleftharpoons ArX_2$, carried out in aqueous solution at 25°C, $\rho_X = \Delta\sigma_I(X_1 \rightarrow X_2)/\Delta\sigma_I(CO_2H \rightarrow CO_2^-)$, which is clearly the basis of the dependence of the magnitude and sign of ρ on charge type and position of reaction site.

For a set of reaction series of the type $RGX_1 \rightleftharpoons RGX_2$, where G is some linking group, e.g. m-C_6H_4, carried out under the same conditions of solvent and temperature, the reaction parameters will be given by $\rho_X^G = \tau_G(\sigma_{X_1} - \sigma_{X_2})$ and similarly for the set $RGY_1 \rightleftharpoons RGY_2$, $\rho_Y^G = \tau_G'(\sigma_{R_1} - \sigma_{Y_2})$. The factors τ_G and τ_G' will only be different if the solvent and temperature conditions of the two sets are different. These results imply the existence of ρ–ρ relationships of the form

$$\frac{\rho_X^G}{\rho_Y^G} = \frac{\tau_G}{\tau_G'} \frac{(\sigma_{X_1} - \sigma_{X_2})}{(\sigma_{Y_1} - \sigma_{Y_2})} \tag{2.29}$$

in which the ratio of the reaction parameters is independent of the linking group. O'Farrell and Miller (1963) and Ritter and Miller (1964) have observed relationships of this type in which the $Y_1 \rightarrow Y_2$ reaction is the dissociation of carboxylic acids in water at 25°C and the $X_1 \rightarrow X_2$ reaction is either the alkaline hydrolysis of ethyl carboxylate in 88% aqueous ethanol at 30°C ($\rho_X^G \simeq 2\cdot4\,\rho_Y^G$) or the reaction of carboxylic acid with diphenyl diazomethane in ethanol at 30°C ($\rho_X^G \simeq 0\cdot5\,\rho_Y^G$).

The Hammett equation fails when the reaction site is not in a side chain and is not therefore applicable to nuclear substitution reactions. In a sense the reactions of anilines, phenols and their derivatives are in this class, since the N and O atoms of these groups are involved with the aromatic group in an extended π-electron system. When the substituent has little or no effect upon the change in conjugative interaction between the aromatic group and the reaction site, then its σ value should still give a quantitative account of the substituent effect. That this is so for many m-substituents is indicated by the correlations given in Table 2.2.

Brown and Nelson (1953), Okamoto and Brown (1957) and Brown and Stock (1962) have examined the applicability of LFER to electrophilic aromatic substitution. For a given substrate in a series of reactions, a relationship is observed between the "activity" of the reagent and its "selectivity" that can be expressed by equation 2.33.

$$\log p_f = c \log(p_f/m_f) \tag{2.33}$$

where p_f and m_f are the partial rate factors for p- and m-substitution, respectively, and c depends upon the substrate. The partial rate factors are simply relative rate constants determined from isomer distribution and product composition experiments, i.e. $p_f = k_p/k_0$ and $m_f = k_m/k_0$, where k_0 applies to benzene itself. Equation 2.33 in reality correlates two types of selectivity, $p_f/m_f = k_p/k_m$ between positions and p_f between substrates.

If an equation of the Hammett type applies to electrophilic aromatic substitution such that

$$\log(k_p/k_0) = \log p_f = \sigma_p^+ \rho^+ \text{ etc.} \tag{2.34}$$

then the constant, c, in equation 2.33 will be given by

$$c = \sigma_p^+/(\sigma_p^+ - \sigma_m^+) \tag{2.35}$$

In the early treatment it was assumed that for the CH_3 group $\sigma_m^+ = \sigma_m$ so that from c value for toluene σ_p^+ could be calculated, and hence ρ^+ values. Reasonably successful correlations were then obtained employing σ^+ values of other substituents.

Subsequently it was reasoned that solvolysis of the substituted phenyl-dimethylcarbinyl (cumyl) chlorides, reaction series 33 of Table 2.2, might serve as a suitable "defining" reaction series for σ^+ values, since the transition state, having considerable carbonium ion character, should resemble the electrophilic-substitution transition states. Several series were successfully correlated by means of the σ^+ values, although in several cases this parameter either overestimates or underestimates the enhancement of the substituent effect.

As indicated, above, the σ'' and σ^0 treatments suggest that the abnormalities in substituent effects cannot be accommodated by a single σ^+ value or σ^- value for each substituent. Indeed extensions of Brown's approach have led to various σ^+ scales. Yukawa and Tsuno (1959) suggest that for each substituent the inductive contribution probably remains constant and hence that

$$\sigma^+ - \sigma^0 = \sigma_R^+ - \sigma_R^0 = r\Delta\sigma_R^+ \tag{2.36}$$

where r depends upon the reaction series, i.e. it is an additional reaction parameter, and is set equal to unity for the solvolysis of the cumyl chlorides. This makes $\Delta\sigma_R^+$ simply the difference between Brown's σ^+ value and the σ^0 value. A modified Hammett equation results, viz.

$$\log(k/k_0) = \rho(\sigma + r\Delta\sigma_R^+) \tag{2.37}$$

which gives a good account of substituent effects on electrophilic reactions. Equation 2.37 will inevitably correlate the data better than the original Hammett equation, with or without the σ^+ values, since an additional adjustable parameter, r, has been introduced.

In the σ_I, σ_R treatment equation 2.37 is equivalent to

$$\log(k/k_0) = \rho_I\sigma_I + \rho_R\sigma_R \tag{2.38}$$

in which $\rho_I = \rho$, $\rho_R = (1-r)\rho_I$ and $\sigma_R = \sigma_R^0 + \sigma_R^+ r/(1-r)$

A similar treatment can of course be applied to nucleophilic substitution reactions and to reactions of anilines and phenols employing equations of the same form as equation 2.34 containing σ^- and ρ^-, or as equation 2.37 containing some suitably based $\Delta\sigma_R^-$. Table 2.8 indicates the magnitude of the enhanced substituent parameters required in the simple treatments.

TABLE 2.8. ENHANCED SUBSTITUENT PARAMETERS

Substituent	σ^n	σ^+	Substituent	σ^n	σ^-
p-N(CH$_3$)$_2$	-0.17	-1.7	p-N$_2^+$	(2.0)	3.0[a]
					2.0[b]
p-OCH$_3$	-0.11	-0.80	..	..	
m-OCH$_3$	0.08	0.05	p-N(CH$_3$)$_3^+$	0.80	0.7[a]
					1.1[b]
p-CH$_3$	-0.13	-0.32	..	..	..
p-F	0.06	-0.07	p-NO$_2$	0.78	1.24
p-Cl	0.24	0.11	p-CN	0.67	0.92

a. For reactions of anilines and phenols. b. For nucleophilic aromatic substitutions.

The σ^+ and σ^- parameters have very little predictive value. Probably the best method of assessing relative-reactivity data, where nuclear substitutions are believed to be involved, follows the σ^n or σ^0 treatments. In this way a ρ value is established based on "normal" substituent behaviour, i.e. m-substituents. The magnitude and sign of ρ will be indicative of the reaction process taking place, and the extent of abnormality shown by p-substituents can be used to assess the extent of direct conjugative interaction.

All aromatic systems, including heterocycles, being planar and rigid, have substituent reaction-centre dispositions like the m- and p-positions in benzene. The Hammett equation should be applicable to these systems by using the appropriate unsubstituted derivative although new sets of substituent and reaction parameters are required.

After benzene, the next simplest aromatic system is naphthalene. For each of the two possible positions, α or β, of the reacting side chain there are five remote substituent positions. These may be labelled 3α, 4α, 5α, 6α, 7α and 4β, 5β, 6β, 7β, 8β, respectively. Although there is some additional complexity, there is a fivefold increase in information to be obtained from studies of substituted naphthalene derivatives over that obtainable from benzene derivatives, i.e. m and p only. This provides the opportunity of identifying more than two modes of substituent effect and of setting more critical tests of theoretical models.

Using results for α-naphthalene series, Dewar and Grisdale subjected F and M values of equation 2.8 to test. A similar test for β-naphthalene series has also been reported by Wells and Adcock (1965). The performance of this simple treatment is quite good, but some of its deficiencies are evident. In particular the term F/r_{ij} proves inadequate in the case of 7α- and especially 8β-substituent effects. In all other cases the neglect of dielectric constant has little effect upon the correlation, since it will be essentially the same for all positions. Also factors of the type $1/r_{ij}$ and $\cos\theta_{ij}/R_{ij}^2$ are approximately proportional to one another, so that the former will prove suitable even if the latter should have been employed. (θ_{ij} is the angle between the substituent group dipole vector and the line of length R_{ij} between its centre and the reaction site.) The effective dielectric constant in the 7α- and 8β-cases is likely to be larger owing to the inclusion of solvent molecules into the intramolecular space. This has the effect of reducing direct electrostatic interactions, so that the observed effects of dipolar substituents in these positions is smaller than calculated by F/r_{ij}. The 8β-effect is observed to be even smaller than 7α, presumably owing to the fact that θ_{ij}, unlike all other cases, including m and p, is now large, so that $\cos\theta_{ij}$ is small.

There appears to be a further oversimplification in the F–M treatment, in that the π_{ij} term implies a greater differentiation between conjugating and non-conjugating positions than is observed. The results suggest a substantial secondary relay of π-electronic effects that is not accounted for by π_{ij}.

When all the available naphthalene-reactivity data is examined in terms of the σ_I and σ_R^0 parameters, which it is hoped are independent of substituent positions, one obtains the results summarized in Table 2.9.

It can be seen immediately that there is a considerable transmittance of substituent effects from one ring to the other both in the "inductive" and the "resonance" senses and although there are resemblances between m, and 3α and 4β, and between p and 4α, these positions are more different than might at first sight have been expected. This is particularly evident in the ρ_R values.

The unusual behaviour of the 7α and 8β positions referred to above and ascribed to direct electrostatic interactions is evident in the ρ_I values.

Excluding the 5α result, which seems unusually large, in reaction series A an average $\rho_I = 0\cdot81$ will accommodate the 6α, 5β, 6β and 7β data. For reaction series C, $\rho_I = 1\cdot36$ will fit both 6α and 7β and the ratio $1\cdot36/0\cdot81 = 1\cdot68$ is not very different from the 5α ratio, i.e. $1\cdot6/1\cdot1 = 1\cdot45$. The 7α value on the other hand is no longer small.

TABLE 2.9. CORRELATION OF NAPHTHALENE DATA BY σ_I AND σ_R^0

Position	ρ_I				ρ_R			
	A	B	C	D	A	B	C	D
3α	1·36	2·20	2·86	2·35	0·65	0·86	0·24	0·27
4α	1·58	2·04	b	b	2·38	2·39	b	b
5α	1·11	..	1·59	..	0·69	..	1·01	..
6α	0·79	..	1·34	..	0·69	..	0·88	..
7α	0·64	..	1·66	..	0·75	..	1·56	..
4β	1·23	..	2·30	2·06	0·75	..	1·30	1·44
5β	0·83	..	..	..	0·58	..	..	..
6β	0·83	1·7[a]	..	..	0·88	1·9[a]	..	..
7β	0·79	1·4[a]	1·36	..	0·58	1·1[a]	1·00	..
8β	0·50	..	..	..	0·73	..	..	..
m	1·51	2·47	2·89	2·26	0·51	1·32	0·88	0·39
p	1·46	2·60	b	b	1·77	3·25	b	b

Reaction series. Number in Table 2.2. A—2; B—23; C—16; D—11.
a. Methyl ester saponification. b. Excluded since direct conjugation can take place.

The ρ_R values show some indication of grouping into conjugating and non-conjugating positions. Thus for series A, $6\beta = 8\beta > 7\alpha > 6\alpha \simeq 5\alpha > 7\beta \simeq 5\beta$, and for series C, $7\alpha > 5\alpha \simeq 7\beta > 6\alpha$. Again the 5α value appears unusual.

Derivatives of pyridine and the quinolines may be regarded as modified derivatives of benzene and naphthalene. In particular, the basicity of the ring nitrogen corresponds to a "reacting side chain" composed of the ring atom alone. Correlation of the pK values in water at 25°C yields

3-Substituted pyridines (m) $pK = 5·88\sigma_I + 2·58\sigma_R^0$

3-Substituted quinolines (3α) $pK = 5·44\sigma_I + 1·46\sigma_R^0$

4-Substituted isoquinolines (4β) $pK = 5·54\sigma_I + 2·54\sigma_R^0$

The 4-substituted pyridines and quinolines are not well correlated by the σ^o nor σ^n values, but the deviations are not of the type that can be accommodated by the σ^+ values or the Yukawa–Tsuno treatment. Evidently the ring-nitrogen protonation is not a good model for electrophilic nuclear substitution.

Some success has been achieved in attempts to assess the polar effects of o-substituents and derive σ_o values. It is not yet clear to what extent these parameters are quantitative and whether they are on the same scale of the σ_m and σ_p values. Further, several unexplained peculiarities have arisen in the treatments.

Taft's method of evaluating polar effects in aliphatic substrates (see below) has been applied to o-substituent effects on benzoate ester hydrolyses, yielding the parameters listed in the third column of Table 2.10. Although the method yields substituent parameters relative to the CH_3 group, adjustment to $\sigma_H = 0.00$ can be made. The scale is set by assuming $\rho_o = \rho$ for the ester hydrolyses. These parameters can be employed successfully to correlate a few other reaction series including those corresponding to series 1 ($\rho_o = 1.79$), 14 ($\rho_o = 2.90$) and 36 ($\rho_o = -2.66$) of Table 2.2. Unfortunately, the corresponding reaction parameters in Table 2.2 are $\rho_1 = 1.00$, $\rho_{14} = 2.94$ and $\rho_{36} = -3.21$, and it is not clear what these observations mean in terms of the relative sensitivities of these reaction series to substituent polar effects. A whole new set of ρ_o values may be required to place the σ_o values on the same scale of σ_p values. Since these are at present unknown, comparisons between these substituent parameters cannot be made. Nevertheless, with the exception of F, the values do appear to run parallel suggesting the "true" σ_o values may be linearly related to the σ_p values. Inherent in Taft's method is the assumption that all o-substituents have steric effects essentially the same as that of the CH_3 group. This may be approximately true of all the substituents listed in Table 2.10 which are of comparable size, although F must be an exception.

Jones and Smith (1964) have suggested that the vapour-phase intramolecular elimination of o-substituted isopropyl benzoates (yielding propene) may depend upon polar effects alone if interference with solvation is the prime contribution to proximity effects. The fourth column in Table 2.10 lists substituent parameters, σ_o (Jones–Smith), obtained in this way.

Farthing and Nam (1958) have *defined* the polar contribution to the strengths of the o-substituted benzoic acids as equal to σ_p, and set the residual effect, i.e. $-\Delta pK - \sigma_p$, equal to a "steric" parameter σ_s. Forty-five

TABLE 2.10. o-SUBSTITUENT PARAMETERS (RELATIVE TO H)

Substituent	σ_p^n	σ_0 (Taft)	σ_0(Jones–Smith)
OCH_3	-0.27†	-0.39	-0.53
CH_3	-0.13	-0.17	-0.16
F	0.06	0.24	0.16
Cl	0.24	0.20	0.31
Br	0.27	0.21	..
I	0.30	0.21	..
NO_2	0.78	0.80	0.94

† $-\Delta pK_a$ for p-$CH_3OC_6H_4CO_2H$ in water at 25°C.

reaction series were examined by means of equation 2.39

$$\log(k/k_0) = \rho_E\sigma_p + \rho_s\sigma_s \tag{2.39}$$

many giving excellent correlations. A plot of the ρ_E values against the usual ρ values is reported to be fairly linear, but, as in this case of most four-parameter treatments, the correlations may be deceptively good. The $\rho_s\sigma_s$ term may simply serve as a flexible correction factor for the inadequacies of $\rho_E\sigma_p$.

Charton (1960) has examined several reaction series in which o-effects are well represented by σ_p. These are mainly acid dissociation and ester hydrolyses in which the functional group is separated from the aromatic system by two-atom side chains, such as CH_2CH_2, $CH=CH$, OCH_2, SCH_2 and $SeCH_2$. Substituent effects appear to be some 75% smaller than p-substituent effects.

A few reaction series have been examined in which the substrates contain a constant o-substituent in addition to the variable substituent. In benzoic acid dissociations o-NO_2 and o-Cl appear to lower the ρ value, whereas o-CH_3 and o-OH appear to raise ρ. These observations cast doubt upon the above methods of derivation of σ_o values and raise the general question of the effect of multiple variations in LFER.

Equation 1.21 can be expressed for the effect of two substituents X and Y as

$$\log(k_{XY}/k_{00}) = \rho(\sigma_X + \sigma_Y) + q\sigma_X\sigma_Y \tag{2.40}$$

which indicates that substituent effects are only additive when $q = 0$.

The additional term in equation 2.40 arises as follows

$$\log(k_{XY}/k_{00}) = \log(k_{XY}/k_{X0}) + \log(k_{X0}/k_{00})$$

$$= \sigma_Y\rho_X + \sigma_X\rho \tag{2.41}$$

$$= \log(k_{XY}/k_{0Y}) - \log(k_{0Y}/k_{00})$$

$$= \sigma_X\rho_Y + \sigma_Y\rho \tag{2.42}$$

where ρ_X and ρ_Y are reaction parameters appropriate to the series with constant substituents X and Y, respectively.

It follows from equations 2.41 and 2.42 that

$$(\rho_X - \rho)/\sigma_X = (\rho_Y - \rho)/\sigma_Y = q \tag{2.43}$$

where q is independent of X and Y.

It is certainly true for o-substituents that $\rho_X \neq \rho$ and therefore that $q \neq 0$. In fact, since the σ_0 values of NO_2 and Cl, on the one hand, are probably positive, while those for CH_3 and OH are negative, q appears to be negative.

For many predictive purposes, it is probably a fairly good approximation to assume addivity of substituent effects, provided there are no obvious steric or resonance interactions. However the only dihydroxybenzoic acid whose pK value can be accurately represented by additive hydroxy effects is the 2,4-isomer. The 3,5-isomer has a ΔpK value some 30% lower than predicted, but the dichlorobenzoic acids show almost additive effects.

When substituent effects on the reagent and substrate can be separately correlated by the Hammett equation, then the combined data may show an essentially additive correlation if the appropriate q factor is close to zero, i.e. each ρ value is almost unaffected by the constant substituent. One case is the reaction between substituted diphenyldiazomethanes (YDDM) and substituted benzoic acids ($XC_6H_4CO_2H$) for which

$$DDM + XC_6H_4CO_2H \quad \log k = 2 \cdot 51\sigma_X - 0 \cdot 14$$

$$YDDM + C_6H_5CO_2H \qquad = 1 \cdot 57\sigma_Y - 0 \cdot 10$$

$$YDDM + XC_6H_4CO_2H \qquad = 2 \cdot 38\sigma_X - 1 \cdot 62\sigma_Y - 0 \cdot 11$$

Multiple variations involving solvent and reagent changes are discussed in Chapters 4 and 5.

Considerable use has been made of Hammett correlations in the study of reaction mechanisms. It is almost a standard practice, when possible, to measure rates for several aromatic substrates either as an indicator of the type of process involved or as a test of a postulated mechanism. The recommended procedure following current examinations of the Hammett equation is to base a ρ value on the observed $\log(k/k_0)$ corresponding to the "well behaved" substituents, i.e. mainly *meta*. The magnitude and sign of ρ will then be indicative of the site of reaction and the difference in its ionic character in the initial and transition states. Whenever the reaction involves changes in conjugative interactions with the aromatic system certain p-substituents will behave abnormally as judged by deviations from σ^n or σ^0. The resonance interaction can be estimated quantitatively.

The procedure is relatively straightforward when the reaction parameters are large, e.g. $|\rho| > 1$. However a small ρ value may be ambiguous, especially when the correlation is poor and p-substituents show signs of abnormality. The reaction cannot be identified as one without ionic character on this basis alone. Thus the observed rate process may depend upon several steps, each with a substantial substituent dependence but of opposite sign. A simple case is that when the observed rate arises from the rate-controlling reaction of the equilibrium concentration of the effective substrate. In this case

$$\log(k/k_0)_{obs} = \log(K/K_0)_{eq} + \log(k/k_0)_{rate}$$

$$= \sigma(\rho_{eq} + \rho_{rate}) \qquad (2.44)$$

for which ρ_{eq} and ρ_{rate} might be approximately equal but of opposite sign. In many reactions the observed rate depends upon several steps, some of which may be reversible and a common mechanism yields expressions of the form

$$k_{obs} = k_1/(1 + k_2/k_3) \tag{2.45}$$

In such cases steps 1, 2 and 3 may individually follow the Hammett equation with high precision and yet poor correlations be obtained using k_{obs}.

As a general rule most polar reaction series give satisfactory correlations with the substituent parameters. On the other hand free-radical reaction series may be correlated by the σ^n or σ^0 parameters, or occasionally by the σ^+ parameters, but frequently any substituent irrespective of its polarity increases radical stability and increases the rate of radical formation.

2.2 THE TAFT EQUATION

If the logarithms of the rates of hydrolysis of substituted aliphatic esters are plotted against the pK_a values of the corresponding acids, there is no indication of any relationship between the effect of substituents in the two reactions series. A similar situation applies in the case of o-substituted aromatic substrates. Clearly the prime condition for the observation of the Hammett correlation, namely the absence of proximity effects, no longer holds. Linear free-energy correlations may still be observed, provided that severe restrictions are placed upon the type of reactions taking place. Thus the Brønsted equation (see Chapter 5) may apply if the changes are in a limited class of acid- and base-catalysed reactions.

Regular trends amongst the acid strengths have been noted on several occasions. This implies that it is largely the ester-hydrolysis data that are responsible for the failure to find simple correlations. Following a suggestion by Ingold (1930), Taft (1952, 1953) examined the *difference* between the substituent effects on base-catalysed and acid-catalysed esterification and ester hydrolysis. Hammett reaction parameters for several reactions of these types are listed in Table 2.11. It is clear that base-catalysed ester reactions show a very substantial response to substituent polar effects whereas the acid-catalysed reactions are relatively insensitive.

On examination of the postulated transition-state configurations

$$
\left[
\begin{array}{c}
\text{OH} \\
\vert\vert \\
\text{R}-\text{C}\cdots\text{OH}_2 \\
\vert \\
\text{OR}'
\end{array}
\right]^{+}
\qquad
\left[
\begin{array}{c}
\text{O} \\
\vert\vert \\
\text{R}-\text{C}\cdots\text{OH} \\
\vert \\
\text{OR}'
\end{array}
\right]^{-}
$$

$$A \qquad\qquad\qquad B$$

TABLE 2.11. HAMMETT REACTION PARAMETERS FOR ESTER REACTIONS

Base-catalysed reaction		ρ_B
Hyd. $ArCO_2C_2H_5$; 85% EtOH;	25°	2·55[a]
Hyd. $ArCO_2C_2H_5$; 88% EtOH;	30°	2·55[a]
Hyd. $ArCO_2C_2H_5$; 60% Acetone;	25°	2·47[a]
Hyd. $ArCO_2CH_3$; 60% Acetone;	25°	2·37[a]
$ArCO_2$(−)-Menth+CH_3O^-; CH_3OH;	40°	2·62[a]
Mean $\rho_B = 2·47 \pm 0·04$[c]		
Hyd. $ArCH_2OAc$; 60% Acetone;	25°	0·74[a]
Hyd. $ArCONH_2$; H_2O;	100°	1·07[b]
Hyd. $ArCONH_2$; 60% EtOH;	65°	1·27[b]

Acid-catalysed reaction		ρ_A
Hyd. $ArCO_2C_2H_5$; 60% EtOH;	100°	0·14[b]
Hyd. $ArCO_2C_2H_5$; 60% Acetone;	100°	0·11[b]
$ArCO_2H+C_6H_{11}OH$;	65°	0·52[b]
$ArCO_2H+CH_3OH$;	25°	−0·52[a]
Mean $\rho_A \simeq 0·1 \pm 0·2$[c]		
Hyd. $ArCH_2OAc$; 60% Acetone;	25°	−0·05[b]
Hyd. $ArCONH_2$; H_2O;	100°	0·12[b]
Hyd. $ArCONH_2$; 60% EtOH;	65°	−0·31[b]

a. See Table 2.2. b. Jaffé (1953). c. All ρ values adjusted to 25°C.

it seems reasonable to assume that differences in steric effects and conjugative effects between the initial and transition states will be closely similar for otherwise identical acid- (A) and base-catalysed (B) reactions. If this is so, then it may be possible to express the substituent effect on the base-catalysed reaction by

$$\log(k/k_0)_B = P + S + R \qquad (2.46)$$

where P, S and R are, respectively, polar, steric and resonance contributions, and, similarly, the substituent effect on the corresponding acid-catalysed reaction by

$$\log(k/k_0)_A = S + R \qquad (2.47)$$

where the *same* S and R contributions are involved and the polar contribution is negligible. The combination of equations 2.46 and 2.47 enabled Taft to separate the polar effect and set up equation 2.48

$$\log(k/k_0)_B - \log(k/k_0)_A = P \equiv \sigma^* \rho^* \qquad (2.48)$$

where σ^* is a substituent parameter and ρ^* is a reaction parameter.

In an attempt to place the σ^* values on the same scale as the Hammett σ values, the ρ^* values for alkyl ester hydrolysis was set equal to 2·48 (cf. the mean values of ρ_B and ρ_A given in Table 2.11). With these ρ^* values, and average values of $\log(k/k_0)_B$ and $\log(k/k_0)_A$ for several alkyl groups and solvent compositions, equation 2.48 becomes a *definition* of the σ^* values.

The performance of the σ^* parameters in the correlation of reactivity data is illustrated in Table 2.12. These may be regarded as tests of the assumption that the polar effect has been separated from other effects and that the σ^* values are useful quantitative measures of this effect. In all cases the reference substituent is the CH_3 group and not hydrogen. Except for the standard reaction series, the Taft equation is employed in the form of equation 2.49

$$\log(k/k_0) = \sigma^*\rho^* \tag{2.49}$$

TABLE 2.12. CORRELATIONS BY THE σ^* VALUES (TAFT, 1956)

Reaction	Conditions	n^a	ρ^*	s^b	$\log k_0^c$
1. $R.CO_2C_2H_5$, hyd.†	Various; 25°	..	$-2\cdot48$	..	..
2. $R.CO_2H \rightleftarrows RCO_2^-$	H_2O; 25°	16	$1\cdot72\pm0\cdot03$	$0\cdot06$	$-4\cdot65$
3. $RCH_2CH(OEt)_2$, acid hyd.	50% dioxan; 25°	7	$-3\cdot65\pm0\cdot09$	$0\cdot08$	$-0\cdot73$
4. $R.CO_2H+(C_6H_5)_2CN_2$	EtOH; 25°	12	$1\cdot18\pm0\cdot04$	$0\cdot06$	$-1\cdot94$
5. $RCH_2OH \rightleftarrows RCH_2O^-$	i-PrOH; 27°	8	$1\cdot36\pm0\cdot09$	$0\cdot09$	$-0\cdot07$
6. $CH_3CH(OH)_2$ $-H_2O(RCO_2H)\ddagger$	Acetone; 25°	13	$0\cdot80\pm0\cdot02$	$0\cdot02$	$-0\cdot11$
7. $CH_3COCH_3+I_2(RCO_2H)$†	H_2O; 25°	5	$1\cdot14\pm0\cdot02$	$0\cdot02$	$-7\cdot65$
8. NH_2NO_2, decomp. $(RCO_2^-)\ddagger$	H_2O; 25°	7	$-1\cdot43\pm0\cdot04$	$0\cdot07$	$-0\cdot37$
9. $R_1R_2R_2CCl$, solvolysis$\ddagger$§	80% EtOH; 25°	13	$-3\cdot29$	$0\cdot22$	..
10. $RCH_2OSO_2C_7H_7$, solvolysis	EtOH; 100°	6	$-0\cdot74$	$0\cdot03$	..
11. $RCH_2Br+C_6H_5S^-$	MeOH; 20°	5	$-0\cdot61$	$0\cdot02$	..

Reactions of o-substituted aromatic substrates ($R_0 = CH_3$)

12. $ArCO_2C_2H_5$, hyd.†	Various; 25°	..	$2\cdot48$	..	..
13. $ArCO_2H \rightleftarrows ArCO_2^-$	H_2O; 25°	7	$1\cdot79\pm0\cdot13$	$0\cdot15$	$-3\cdot69$
14. $ArNH_3^+ \rightleftarrows ArNH_2$	H_2O; 25°	5	$2\cdot89\pm0\cdot15$	$0\cdot19$	$0\cdot61$
15. $CH_3CH(OH)_2$ $-H_2O(ArCO_2H)$†	Acetone; 25°	5	$0\cdot77\pm0\cdot02$	$0\cdot02$	$0\cdot34$
16. $ArNH_2+C_6H_5COCl$	Benzene; 25°	4	$2\cdot66\pm0\cdot22$	$0\cdot28$	$-1\cdot86$

a. Number of substituents employed. b. Probable error of fit of a single observation. c. Intercept at $\sigma^* = 0$, k in $\sec^{-1}$ for rate series.
† $\log(k/k_0)_B - \log(k/k_0)_A$.
‡ General acid catalysis. Reaction series 2 and 13 requires that all reaction series following the Brønsted equation will also follow the Taft equation.
§ Correlation with $\Sigma\sigma^*$ with $(CH_3)_3CCl$ as reference.

The original list of σ^* values was composed largely of those for alkyl groups, but later compilations, based upon all available data, contain many substituents having more obvious polar character. A selection of values is given in Table 2.13. Parameters for o-substituted aromatic substrates can be found in Table 2.10.

Further ρ^* values for correlations of acid dissociation in water at 25°C are given in Table 2.14.

Ritchie (1961) has employed the arguments of Hine (1959) in the case of substrates of the type $R(CH_2)_n X$ and, corresponding to equations 2.30 and 2.31, has shown that

$$\rho^*_{CH_2X} = \tau^*_{CH_2}(\sigma^*_{X_1} - \sigma^*_{X_2}) \tag{2.50}$$

and that

$$\frac{\rho^*_{CH_2CH_2X}}{\rho^*_{CH_2X}} = \frac{(\sigma^*_{CH_2R_1} - \sigma^*_{CH_2R_2})}{(\sigma^*_{R_1} - \sigma^*_{R_2})} = \alpha^*_{CH_2} \tag{2.51}$$

where $\tau^*_{CH_2}$ is independent of X_1 and X_2 being determined by the reaction conditions and overall substrate structure. The "relay factor" $\alpha^*_{CH_2} = \tau^*_{CH_2CH_2}/\tau^*_{CH_2}$, should be independent of R, X and the reaction conditions.

TABLE 2.13. TAFT SUBSTITUENT PARAMETERS (SUBSTITUENT R IN RCH_2)

R	σ^*	R	σ^*	R	σ^*
Me_3N^+	2·00	$OCOCH_3$	0·89	NMe_2	0·22
NO_2	1·40	OCH_3	0·66	C_6H_5	0·22
CH_3SO_2	1·38	CO_2R	0·66	$CH=CH_2$	0·12
CH_3SO	1·33	$COCH_3$	0·62	$C_6H_4CH_2$	0·08
CN	1·25	NHAc	0·60	H	0·00
F	1·10	OH	0·55	CH_3	−0·10
Cl	1·05	SH	0·47	C_2H_5	−0·12
Br	1·02	SMe	0·42	$i\text{-}C_3H_7$	−0·13
CF_3	0·92	NH_2	0·40	$i\text{-}C_4H_9$	−0·17
I	0·88	O^-	0·27	$SiMe_3$	−0·25

TABLE 2.14. REACTION PARAMETERS FOR ACID DISSOCIATION (H_2O; 25°)

Acid	ρ^*	Acid	ρ^*	Acid	ρ^*
RCH_2CO_2H	1·75[a] (1·72[b])	$RCH_2NH_3^+$	3·80[a] (3·14[b])	$RCH_2PH_3^+$	(2·64[b])
$RCH_2PO_3H_2$	1·16[a]	$(RCH_2)_2NH_2^+$	3·90[a] (3·23[b])	$(RCH_2)_2PH_2^+$	(2·61[b])
RCH_2OH	3·47[a]	$(RCH_2)_3NH_3^+$	4·29[a] (3·30[b])	$(RCH_2)_3PH^+$	(2·67[b])
RCH_2SH	3·73[a]				

a. Based upon more extensive σ^* list. b. Based upon original σ^* values.

The multiple substitution situation has also been examined by Ritchie (cf. equations 2.40 and 2.43), yielding for a variable substituent Y in RYCHX

$$\log(K_{RY}/K_{00}) = (\sigma_R^* + \sigma_Y^*)\rho_{CH_2X}^* + q^*\sigma_R^*\sigma_Y^* \tag{2.52}$$

or a constant substituent Y

$$(\rho_{CHYX}^* - \rho_{CH_2X}^*) = q^*\sigma_Y^* \tag{2.53}$$

In particular for the substrates RCH_2X, R_2CHX and R_3CX one obtains,

$$\log(K_R/K_0) = \sigma_R^*\rho_{CH_2X}^*$$

$$\log(K_{R_2}/K_0) = \sigma_{R_2}^*\rho_{CH_2X}^* = 2\sigma_R^*\rho_{CH_2X}^* + q^*(\sigma_R^*)^2$$

$$\log(K_{R_3}/K_0) = \sigma_{R_3}^*\rho_{CH_2X}^* = 3\sigma_R^*\rho_{CH_3X}^* + 3q^*(\sigma_R^*)^2 \tag{2.54}$$

The substituent parameters conform reasonably well to equation 2.51 provided that the substituent has pronounced polar character. The alkyl groups tend to deviate from this correlation suggesting that the original list of σ^* values for these groups is in error. It appears that most alkyl groups have essentially the same small polar effect and that perhaps inadequacies in the separation of polar and steric effects are responsible for the reported σ^* values.

For dipolar substituents, it is found that for aqueous solution at 25°C

$$\Delta pK_{RCH_2CH_2CO_2H} = 0.40\,\Delta pK_{RCH_2CO_2H} \tag{2.55}$$

and

$$\Delta pK_{RCH_2CH_2NH_3^+} \simeq 0.43\,\Delta pK_{RCH_2NH_3^+} \tag{2.56}$$

Slightly different relay factors are obtained for different reaction series so that equation 2.51 does not hold *exactly*. The relay factors and their differences can be accounted for on the basis of an electrostatic model for aliphatic substituent effects (see Chapter 3). Nevertheless, $\sigma_R^* = 2.5\,\sigma_{CH_2R}^*$ provides a good approximation for calculating new substituent parameters.

Based largely upon the strengths of the substituted acetic acids, it is found that halogen-containing substituents are in the ratio

$$F_3:F_2:F = 2.16:1.57:1$$

$$Cl_3:Cl_2:Cl = 2.15:1.80:1$$

$$Br_3:Br_2:Br = 2.14:1.70:1 \tag{2.57}$$

which, from equations 2.54, requires that

$$q^* \simeq -0.3\rho_{CH_2X}^* \tag{2.58}$$

Although the ρ^* value for reaction series 1 of Table 2.13 has been set equal to the corresponding ρ values, this does not necessarily place the σ^* values on the same scale as the σ values. The σ^* values could have been defined by reaction series 2 employing the corresponding ρ value, which, by definition, is unity. In this way an equally good set of substituent parameters would have been obtained, all 1·7 times larger than those of Table 2.14. All the ρ^* values would then be correspondingly 1·7 times smaller. This appears at first sight to imply that one or other ρ^* value is abnormal.

Substrates of the type $RC_6H_4CH_2X$ are however suitable for treatment as substituted aromatic or aliphatic compounds, and a common scaling of the Hammett and Taft parameters can be achieved employing the identity

$$(\sigma^*_{R_1C_6H_4} - \sigma^*_{R_2C_6H_4}) \equiv (\sigma_{R_1C_6H_4} - \sigma_{R_2C_6H_4}) \tag{2.59}$$

Following the reasoning leading to equations 2.31 and 2.51, one may write for substrates of the type $RC_6H_4.C_6H_4X$

$$\frac{(\sigma_{R_1C_6H_4} - \sigma_{R_2C_6H_4})}{(\sigma_{R_1} - \sigma_{R_2})} = \frac{\rho_{C_6H_4X}}{\rho_X} = \alpha_{C_6H_4} \tag{2.60}$$

and for substrates of the type $RC_2H_4CH_2X$

$$\frac{(\sigma^*_{R_1C_6H_4} - \sigma^*_{R_2C_6H_4})}{(\sigma_{R_1} - \sigma_{R_2})} = \frac{\rho_{CH_2X}}{\rho^*_{CH_2X}} \tag{2.61}$$

It follows from equations 2.59, 2.60 and 2.61 that

$$\rho^*_{CH_2X} = \rho_{CH_2X}/\alpha_{C_6H_4} \tag{2.62}$$

where $\alpha_{C_6H_4}$ is simply the relay factor for the C_6H_4 group.

A fairly reliable value of $\alpha_{C_6H_4}$ can be obtained from the ester hydrolysis data of Berliner and Liang Huang Lin (1953) as $0·608/2·545 = 0·24$, and a very approximate value of $\sim0·3$ can be estimated from the acidity measurements of Berliner and Blommers (1951). Calculated $\rho^*_{CH_2X}$ values are given in Table 2.15.

The assumption that $\rho_b^* = 2·5$ should lead to values for ρ_a^* of $2·1 \times 2·5/3·4 = 1·5$, and for ρ_c^* of $4·0 \times 2·5/3·4 = 3·0$. Although these are

TABLE 2.15. CALCULATED ρ^* VALUES ON HAMMETT SCALE

Reaction	ρ_{CH_2X}	$\rho^*_{CH_2X}$
(a) $CO_2H \rightleftarrows CO_2^-$; H_2O; 25°	0·50	2·1
(b) $CO_2C_2H_5 + OH^-$; 87% aq EtOH; 25°	0·82	3·4
(c) $NH_3^+ \rightleftarrows NH_2$; H_2O; 25°	1·05	4·0

somewhat smaller than the observed values, i.e. 1·72–1·75 and 3·14–3·80 respectively (see Table 2.14) they are sufficiently similar to indicate that the observed ρ^* values are in no way abnormal.

Returning to equation 2.47, it can be seen that in the absence of resonance interactions $\log(k/k_0)_A$ is considered to be determined almost solely by steric factors. Taft has employed this quantity to define steric parameters

$$\log(k/k_0)_A = E_s \qquad (2.63)$$

employing the hydrolysis of o-substituted benzoate esters or acetate esters as the standard reaction series for aromatic and aliphatic substituents, respectively. In the general case where polar effects are also present, equation 2.64 is employed

$$\delta E_s = \log(k/k_0) - \sigma^*\rho^* \qquad (2.64)$$

in which δ is a reaction parameter measuring the sensitivity to steric effects. For example, the acid-catalysed hydrolysis of benzamides can be correlated to yield $\delta = 0.81 \pm 0.03$.

Some substituent steric parameters are listed in Tables 2.16 and 2.17. In the case of o-substituents in aromatic substrates the magnitude of E_s appears reasonable and other evidence implies that the CH_3 and Br substituents have similar size, whereas Cl and I are certainly smaller and larger,

TABLE 2.16. STERIC PARAMETERS FOR o-SUBSTITUTED BENZOATES

Substituent	R_s	Substituent	E_s
C_6H_5	−0·90	OCH_3	0·97
NO_2	−0·71	OC_2H_5	0·86
I	−0·20	Cl	0·18
CH_3	0·00	Br	0·01

TABLE 2.17. STERIC PARAMETERS FOR ALIPHATIC SUBSTRATES

Substituent	E_s	Substituent	E_s
CH_3	0·00	cyclo-C_4H_7	−0·06
C_2H_5	−0·07	cyclo-C_5H_9	−0·51
n-C_3H_7	−0·36	cyclo-C_6H_{11}	−0·79
n-C_4H_9	−0·39	cyclo-C_7H_{13}	−1·10
n-C_5H_{10}	−0·40	..	..
i-C_3H_7	−0·47	CF_3	−1·16
t-C_4H_9	−1·54	CCl_3	−2·06
neo-C_5H_{11}	−1·74	CBr_3	−2·43

respectively. It is not, however, clear what the large positive E_s values for the alkoxy-groups can mean. These substituents are not as small as the steric parameters appear to require, so that some other factor must be involved. The steric parameters may have their greatest utility for hydrocarbon groups where the assumption of negligible polar and resonance contributions is most nearly valid.

The use of equation 2.64 as a four-parameter correlation is illustrated by the base-catalysed transmethylation of methyl esters (Pavelich and Taft, 1957), for which

$$\log(k/k_0) = (2\cdot70 \pm 0\cdot07)\sigma^* + (1\cdot30 \pm 0\cdot06)E_s$$

It must be realized, however, that four-parameter correlations may appear better than they really are and that a modified set of parameters (see below), may perform equally well. In addition, the general nature of multiple correlations, see equations 2.40 and 2.43, suggests a third term $q^*\sigma^*E_s$ should be included to account for variation in ρ^* and δ with steric and polar effects, respectively.

Much reactivity data still cannot be correlated by the Taft equation even when a correcting term for steric effects has been introduced. There can be many reasons for this situation, but in a few cases it is suggested that the deviations may be used as measures of other variables, particularly resonance effects, hyperconjugative interactions and neighbouring-group participation.

In several reaction series, α,β-unsaturated substituents behave abnormally as judged by their σ^* and E_s values. Thus benzoic acid and acrylic acid are found to be weaker than expected. Changes in resonance interactions are implicated and would be measurable if the polar and steric contributions to $\log(k/k_0)$ could be subtracted out accurately.

Some of the difficulties of this approach are illustrated by the case of the hydrolysis of ketals and acetals. Taft and Kreevoy (1957) have noted that α,β-unsaturated substituents cannot be correlated, since evidently resonance interactions are involved, and the scatter in the correlation of other substituents, largely alkyl groups, appears to be related to the number of hydrogens on the α-carbon atoms (n_H). An improved correlation is obtained by using

$$\log(k/k_0) = 3\cdot60\Sigma\sigma^* + 0\cdot54(n_H - 6) \qquad (2.65)$$

in which it appears that a contribution arising from C—H hyperconjugation has been introduced. Further improvement is obtained by introducing a smaller term dependent on the number of α-C—C bonds, i.e. C—C hyperconjugation. Ritchie (1961) has, however, drawn attention to the fact that the σ^* values for the alkyl groups are not at all well known. They may indeed be close to zero, and using this value for all groups without any account for hyperconjugation places these substrates in their correct position with respect to more polar substituted acetals and ketals in terms of hydrolysis rate. It should be noted that the bulk of an alkyl group and hence its steric requirements increases as the number of α-C—H bonds decreases. Hyperconjugative effects, if real, will be distentangled from steric effects only with very great difficulty.

Hancock et al. (1961) conclude that the E_s parameters probably contain a hyperconjugative contribution, and that corrected parameters E_s^c may be obtained from

$$E_s = E_s^c + h(n-3) \tag{2.66}$$

where n is the number of α-hydrogens, three in the case of the standard substituent CH_3, and $h = 0.31$ is estimated by a LCAO—MO method. Significant improvements in correlation by the four-parameter equation 2.64 are reported for the saponification of methyl esters, alkyl acetates and alkyl benzoates. However, in view of the method by which the E_s parameters are obtained, it is not surprising that a better modified set can be derived. It does not follow that because this modification takes into the number of α-hydrogens that the new factor depends upon hyperconjugation.

Very large deviations from the Taft equation are observed if neighbouring-group participation can occur. Streitwieser (1956) has compared k_{obs}/k_{calc} for the acetolysis of some trans-X-cyclohexyl-p-bromobenzene sulphonates. The values 320 for X = $OCOCH_3$, 450 for X = Br and 1.5×10^6 for X = I are clear indications of the rate-enhancing involvement of these substituents. In reaction series 9 of Table 2.12, small but definite indications of steric acceleration have been obtained.

Perhaps the most important feature of aromatic substrates that is responsible for the observation of so many successful correlations by the Hammett equation is their rigidity and geometric character. A number of non-aromatic structures can be envisaged that also possess these features. Two outstanding examples are the 1,4-disubstituted bicyclo[2.2.2]octanes and the trans-1,4-cyclohexanes. Two bulky substituents in trans-1,4-disposition in cyclohexane will lead to an essentially "locked" conformation with both groups equatorial.

Roberts and Moreland (1953) established in the former case that an LFER holds between acid strengths, ester hydrolysis rates and the reaction of acids with diphenyldiazomethane. With reasonable assumptions concerning the value of the "effective" dielectric constant, the relative acid strengths can be accounted for in terms of direct electrostatic interaction between the substituent and the reaction site. A further examination of this system by Holtz and Stock (1964) has substantiated this view. Similar LFER are reported by Siegel and Komarmy (1960) for the trans-4-substituted cyclohexane-1-carboxylic acids. Again the substituent effects can be accounted for on the basis of direct electrostatic effects.

These systems are particularly interesting, since the substituent orientation and distance from the reaction site is closely similar to that in a p-substituted benzene derivative. The linking group in one case is paraffinic and in the other is composed of a π-electron system. Although the ΔpK values for the substituted bicyclo-octanecarboxylic acids and the m-substituted benzoic acids have a similar magnitude; when these are converted to "properly" scaled substituent parameters it is found that the aromatic substituent parameter is some two to three times larger than the aliphatic parameter. The "properly" scaled reaction parameter, ρ', for the bicyclo-octane case can be obtained from ρ_a^* of Table 2.15 together with an allowance for the constant substituents which are essentially three alkyl groups. Thus from equations 2.53 and 2.58

$$\rho' \simeq \rho_a^* + q^* \Sigma \sigma_{CH_3}^* \simeq 2 \cdot 25$$

which has to be compared with the Hammett ρ value defined as $1 \cdot 00$.

Two further systems that can be profitably studied because of their geometric resemblance to the m-position in benzene are the cis-1,3-disubstituted cyclohexanes and the 1,3-disubstituted adamantanes. Some investigations have been reported, e.g. Stetter and Mayer (1962).

Other potentially fruitful substrate structures are the simple unsaturated non-aromatic systems. Some studies have been carried out by Hine and Bailey (1959) and by Charton and Meislich (1958) on the $trans$-3-substituted acrylic acids, 3-substituted cis-crotonic acids and 2-substituted maleic acids. Satisfactory correlations employing the σ_p values have been described. Charton (1961) also finds that the σ_p values give a good account of substituent effects on the strengths of the acetylene carboxylic acids.

No attempt has yet been made to place substituent parameters for these unsaturated systems on the Hammett or Taft scales, but presumably this can be achieved when more results become available. As in the case of the naphthalene system, these extensions of substituent correlations outside the original benzene system provide important information on the origin of substituent effects and are valuable testing grounds for theoretical treatments.

REFERENCES

Berliner, E., and Blommers, E. A. (1951). *J. Am. chem. Soc.*, **73**, 2479.
Berliner, E., and Liang Huang Lin (1953). *J. Am. chem. Soc.*, **75**, 2417.
Brown, H. C., and Nelson, K. L. (1953). *J. Am. chem. Soc.*, **75**, 6296.
Brown, H. C., and Stock, L. M. (1962). *J. Am. chem. Soc.*, **84**, 3298.
Charton, M. (1960). *Can. J. Chem.*, **38**, 2493.
Charton, M. (1961). *J. org. Chem.*, **26**, 735.
Charton, M., and Meislich, H. (1958). *J. Am. chem. Soc.*, **80**, 5940.

Dewar, M. J. S., and Grisdale, P. J. (1962). *J. Am. chem. Soc.*, **84**, 3539.
Farthing, A. C., and Nam, B. (1958). Chemical Society Symposium on Steric Effects in Conjugated Systems, Hull, England.
Hammett, L. P. (1937). *J. Am. chem. Soc.*, **59**, 96.
Hammett, L. P. (1940). "Physical Organic Chemistry," Chapter 7. McGraw-Hill, New York.
Hancock, G. K., Meyers, E. A., and Yager, B. J. (1961). *J. Am. chem. Soc.*, **83**, 4211.
Hepler, L. G. (1963). *J. Am. chem. Soc.*, **85**, 3089.
Hine, J. (1959). *J. Am. chem. Soc.*, **81**, 1126.
Hine, J., and Bailey, W. C. (1959). *J. Am. chem. Soc.*, **81**, 2075.
Holtz, H. D., and Stock, L. M. (1964). *J. Am. chem. Soc.*, **86**, 5186.
Ingold, C. K. (1930). *J. chem. Soc.*, 1032.
Jaffé, H. H. (1953). *Chem. Rev.*, **53**, 191.
Jaffé, H. H. (1959). *J. Am. chem. Soc.*, **81**, 3020.
Jones, D. A. O., and Smith, G. G. (1964). *J. org. Chem.*, **29**, 3531.
McDaniel, D. H. (1961). *J. org. Chem.*, **26**, 4692.
McGowan, J. G. (1949). *J. Soc. chem. Ind. Lond.*, **68**, 253.
O'Farrell, R. M., and Miller, S. I. (1963). *J. Am. chem. Soc.*, **85**, 2040.
Okamoto, Y., and Brown, H. C. (1957). *J. org. Chem.*, **22**, 485.
Pavelich, W. H., and Taft, R. W. (1957). *J. Am. chem. Soc.*, **79**, 4935.
Ritchie, C. D. (1961). *J. phys. Chem., Ithaca*, **65**, 2091.
Ritter, J. D. S., and Miller, S. I. (1964). *J. Am. chem. Soc.*, **86**, 1507.
Roberts, J. D., and Moreland, W. T. (1953). *J. Am. chem. Soc.*, **75**, 2167.
Roberts, J. L., and Jaffé, H. H. (1959). *J. Am. chem. Soc.*, **81**, 1635.
Siegel, S., and Komarmy, J. M. (1960). *J. Am. chem. Soc.*, **82**, 2547.
Stetter, H., and Mayer, J. (1962). *Chem. Ber.*, **95**, 667.
Streitwieser, A. (1956). *J. Am. chem. Soc.*, **78**, 4935.
Taft, R. W. (1952). *J. Am. chem. Soc.*, **74**, 2729, 3120; (1953), **75**, 4231.
Taft, R. W. (1956). *In* "Steric Effects in Organic Chemistry," (M. S. Newman, ed.), Chapter 13. Wiley, New York.
Taft, R. W. (1960). *J. phys. Chem., Ithaca*, **64**, 1805.
Taft, R. W., and Kreevoy, M. M. (1957). *J. Am. chem. Soc.*, **79**, 4011.
Taft, R. W., and Lewis, I. C. (1959). *J. Am. chem. Soc.*, **81**, 5343.
van Bekkum, H., Verkade, P. E., and Wepster, B. M. (1959). *Rec. trav. chim. Pays-Bas Belge*, **78**, 815.
Yukawa, Y., and Tsuno, Y. (1959). *Bull. chem. Soc. Japan*, **32**, 971.
Wells, P. R., and Adcock, W. (1965). *Aust. J. Chem.*, **18**, 1365.

3

Substituent Effects on Reactivity

3.1 Measures of Substituent Effects

Throughout Chapter 2, substituent effects have been examined in terms of relative reactivities, i.e. $\log(k/k_0)$, and the corresponding relative free-energy changes, $\delta_R \Delta F$. This is not, of course, the only area of chemistry where substituent effects are important. It has, however, provided the most data, both in the sense of yielding information and of demanding explanations. As indicated in Chapter 6, substituent effects on spectroscopic quantities tend to be explained in terms of parameters derived from reactivity studies.

The customary choice for the standard substituent, i.e. k_0, is hydrogen, but this may not always be a good choice. Thus in the Taft equation the methyl group is used. In the measurement of polar effects one would wish to make comparisons with a hypothetical substituent that exerts no polar effect, but is otherwise a "typical" substituent. Remote non-conjugating substituents can be usefully related to hydrogen, although in reality the Hammett equation is concerned with the comparison between phenyl and substituted phenyl groups. Conjugating substituents of the first row donor type might be better related to the R_2C^- substituent where the R groups give the donor carbon atom the same electronegativity as aromatic carbon atoms. Where steric effects of various types may be involved and are hopefully cancelled out in the relative-reactivity comparison, a reference with essentially the same steric requirements but negligible polar character is required.

Empirically various sets of substituent parameters based upon $\log(k/k_0)$ values have been described. The σ_m, σ_p and σ^* values are directly obtained from observed relative reactivities and, within the limited areas where they are essentially constant, they at present constitute the most fundamental sets of parameters. In principal all suitable aliphatic data can be expressed in terms of σ^* (and ρ^*) values placed on the same scale as the aromatic parameters. These in turn could be converted to equivalent σ_I values based upon aliphatic reactivities. Various analyses of the σ_m and σ_p values are possible, and the most successful so far is that in terms of σ_I and σ_R values. However, although the two sets of σ_I values appear to be proportional to one another, they cannot be identical, since they arise from a different blend of contributing factors. It is not therefore possible to account for all substituent effects in terms of two parameters.

It is hoped that continued work in this field will either substantiate the σ_I–σ_R separation and clarify the relationship between the aromatic and aliphatic parameters or lead to the adoption of a more satisfactory analysis. Ultimately the "true" variables will be revealed and their measurement achieved.

In the meantime it appears constructive to examine the way in which LFER can arise given the nature of the factors believed to contribute to the total substituent effect. Relationships can be sought between the reactivity parameters and other quantities that appear to have bearing on relative reactivities. Obvious quantities in this category are bond dipole moments, electronegativities and relative electron densities.

3.2 The Origin of Substituent Polar Effects

From the mere existence of the LFER and the nature of the parameters obtained, much can be inferred concerning the origin of substituent polar effects

$$R_1GX_1 + R_2GX_2 \rightleftharpoons R_1GX_2 + R_2GX_1 \tag{3.1}$$

Equation 3.1 represents a generalized equilibrium whose constant, K_1, measures the relative effect of the substituents R_1 and R_2 on the reaction corresponding to $X_1 \to X_2$. X_2 will be a transition state configuration when reaction rates are compared. G is some linking group. (K_1 may equally be regarded as measuring the relative effect of the substituents X_1 and X_2 on the reaction $R_1 \to R_2$.)

The equilibrium constant, K_1, is usually obtained as the ratio, K_2/K_3, of two observed equilibrium or rate constants corresponding to the reactions

$$R_1GX_1 + Y \rightleftharpoons R_1GX_2 + Z \tag{3.2}$$

$$R_2GX_1 + Y \rightleftharpoons R_2GX_2 + Z \tag{3.3}$$

where Y is some reagent(s) and Z some other product(s) common to both compared equilibria.

In the gas phase or as a non-associated solution in an indifferent solvent, the parts of a substrate molecule such as p-$ClC_6H_4NH_2$ are readily identifiable as R, G and X groups. However, in most solutions each part of the molecule will be involved in some kind of solvation, and R, etc., must represent the substituents and any closely associated solvent molecules. From the manner in which the total free-energy change is analysed it follows that any solvation that remains unchanged for RGX as R and X are modified becomes identified with solvation of G. Any solvation unchanged between RGX_1 and

RGX_2 will be interpreted as part of the RG-group and any change between RGX_1 and RGX_2 as part of the difference between X_1 and X_2 even when the solvent molecules concerned may be bonded to RG. Thus for the reaction formally written as

$$RGNH_3^+ \rightleftharpoons RGNH_2$$

the X groups correspond to $NH_3^+ . n$ Solvent and $NH_2 . n'$ Solvent, respectively, where n and n' include not only those solvent molecules associated with the NH_3^+ and NH_2 groups but also those associated with RG in $RGNH_3^+$ but not in $RGNH_2$, and vice versa.

In terms of relative free-energy change, one may write

$$-2 \cdot 303 RT \log K_1 = \partial_R \Delta F_X = \partial_X \Delta F_R \qquad (3.4)$$

where

$$\partial_R \Delta F_X = \Delta F_X(R_1) - \Delta F_X(R_2) \qquad (3.5)$$

and

$$\partial_X \Delta F_R = \Delta F_R(X_1) - \Delta F_R(X_2) \qquad (3.6)$$

with $\Delta F_X(R)$ and $\Delta F_R(X)$ the free-energy changes for the reactions $RGX_1 \rightarrow RGX_2$ and $XGR_1 \rightarrow XGR_2$, respectively.

If the relative free-energy change is expressed in terms of "internal" contributions and "external" or solvent–solute interaction contributions and the arguments of Hepler (1963) can be applied (see Chapter 2), one is led to expect that

$$\partial_R \Delta F \simeq \partial_R \Delta H_{int} \qquad (3.7)$$

It may be further assumed that

$$\delta_R \Delta H_{int} \simeq \delta_R \Delta E \qquad (3.8)$$

where $\delta_R \Delta E$ is the relative change in internal energy made up of σ and π-bond energies, of electrostatic interaction energies and of non-bonded interaction energies when these are not constant.

Since substituent effects are believed to arise from changes in the above factors, whereas "external" enthalpy and entropy changes are expected to arise in a markedly different manner, then the existence of LFER's demands at least that $\delta_R \Delta F$ is proportional to $\delta_R \Delta H_{int}$. The proportionality "constant" of unity implied in equations 3.7 and 3.8 may be an oversimplification, and part of the variation of LFER parameters with solvent and temperature may arise from changes in this proportionality "constant". For the present, it is necessary to assume that the observable quantity $\delta_R \Delta F$ and the theoretically interesting quantity $\delta_R \Delta E$ are to a useful approximation equal.

It has been found empirically (cf. Skinner and Pilcher, 1963) that heats of formation can be quantitatively analysed in terms of individual bond energies and interaction terms. In the present case, one may express the internal energy of the various species, i.e. E_{RGX}, as the sum of independent terms arising from parts of the structure, e.g. E_R, together with interaction terms thus

$$E_{R_1GX_2} = E_{R_1} + E_G + E_{X_2} + I^{X_2}_{R_1G} + I^{R_1}_{X_2G} + I^{G}_{R_1X_2} \qquad (3.9)$$

where the superscript X on the term I^{X}_{RG} serves to indicate that the interaction between R and G may depend upon the nature of X.

The interactions I^{X}_{RG} and I^{R}_{XG} may be described as *direct* interactions, whereas I^{G}_{RX} is an *indirect* interaction between R and X across and/or through G.

It follows that

$$\partial_R \Delta E_X = E_{R_1GX_2} + E_{R_2GX_1} - E_{R_1GX_1} - E_{R_2GX_2}$$
$$= (I^{X_2}_{R_1G} - I^{X_2}_{R_2G}) - (I^{X_1}_{R_1G} - I^{X_1}_{R_2G})$$
$$+ (I^{R_1}_{X_2G} - I^{R_1}_{X_1G}) - (I^{R_2}_{X_2G} - I^{R_2}_{X_1G})$$
$$+ I^{G}_{R_1X_2} + I^{G}_{R_2X_1} - I^{G}_{R_1X_1} - I^{G}_{R_2X_2} \qquad (3.10)$$

It is convenient at this point to examine the information provided by the LFER to see what may be deduced concerning the nature and relative magnitude of the interaction terms involved in equation 3.10.

When G is an aliphatic linking group such as CH_2 the Taft equation may be applicable. One such case is the relative strengths of the substituted acetic acids, XCH_2CO_2H. These data have been explained in terms of σ bond inductive effects (I_σ) and direct electrostatic effects either as alternative explanations or as two contributing factors.

The qualitative aspects of the I_σ effect were first suggested by Lewis (1916) and are customarily formulated for an electronegative substituent, R, as follows

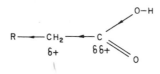

A σ-bond electron shift towards R from the methylene carbon induces a fractional charge on this carbon, and leads to a bond dipole moment. The modified carbon atom withdraws charge from the carboxyl group facilitating proton loss from oxygen, thus strengthening the acid relative to acetic acid.

The effect of the substituent in $RCH_2CH_2CO_2H$ will be smaller, since the induction must traverse an additional σ bond. Waters (1934) has attempted to assess the fall-off factor for induction through a saturated carbon chain from observed polarizability data, and suggests a factor of approximately $\frac{1}{40}$ may be reasonable. When this effect is introduced into molecular orbital calculations a factor of $\frac{1}{10}$ is commonly employed. Empirically the fall-off factor, $\alpha^*_{CH_2}$ for substituent effects in aliphatic substrates is found to be 0·40 so that it appears that the I_σ effect cannot account for the substituent effects that are correlated by the Taft equation.

On the other hand, with reasonable assumptions concerning molecular geometry, bond dipole moments and effective dielectric constant, a good account can be given of relative acid strengths based upon direct electrostatic interactions (cf. Kirkwood and Westheimer, 1938; Bell and Wright, 1961; Holtz and Stock, 1964). In addition, if one compares the electrostatic free energy of interaction between a substituent dipole and the carboxylate anion in RCH_2X and RCH_2CH_2X (fully staggered conformation), and assumes

$$R-\widehat{C}-X = R-\widehat{C}-C = C-\widehat{C}-X = 109\tfrac{1}{2}° \text{ and } r_R = r_X = r_C, \text{ then}$$

$$\alpha^*_{CH_2} \simeq \frac{4(3 - 2\cos 109\tfrac{1}{2}°)(1·25 - \cos 109\tfrac{1}{2}°)^{\frac{3}{2}}}{(0·5 - \cos 109\tfrac{1}{2}°)(13 - 12\cos 109\tfrac{1}{2}°)^{\frac{3}{2}}} = 0·50$$

neglecting differences in effective dielectric constant. Various refinements can be made to this calculation. Thus, the bond lengths are not all equal and in the case of the carboxylate anion, the location of the charged site on the C—X axis may be incorrect. In the comparison of chloroacetate and 3-chloropropionate, $\alpha^*_{CH_2}$ may be some 25% smaller than estimated. This is partially compensated by the neglect of differences in effective dielectric constant, which is probably larger for $XCH_2CH_2CO_2^-$ than for $XCH_2CO_2^-$ by a factor of about 1·7. These two corrections yield $\alpha^*_{CH_2} \simeq 0·40$ which is in exact agreement, although probably fortuitously so, with the observed relay factor.

The conclusion seems clear that a direct electrostatic mechanism for substituent–reaction site interactions is consistent with the observed results whereas σ bond induction is not. The latter is then an unnecessary postulate.

Whether or not σ bond induction plays an important role in substrates of the RX type cannot be determined, largely because of difficulties in assessing the magnitude of relative free-energy changes arising from this and direct electrostatic effects. Nevertheless, for substrates to which the Taft equation is applicable, it seems that equation 3.10 may be drastically simplified to contain indirect interaction terms alone, i.e.

$$\partial_R \Delta E_X \simeq I^G_{R_1 X_2} + I^G_{R_2 X_1} - I^G_{R_1 X_1} - I^G_{R_2 X_2} \qquad (3.11)$$

and that these arise from direct electrostatic interactions. For dipolar substituents, equation 3.11 may be written for X_2, a site bearing a charge ze, and X_1, a dipolar group,

$$\partial_R \Delta E_X \simeq 1{\cdot}44 \times 10^{13} ze \left\{ \frac{\mu_1 \cos \theta_1}{D_1 r_1^2} - \frac{\mu_2 \cos \theta_2}{D_2 r_2^2} \right\} \qquad \text{kcal. mole}^{-1} \qquad (3.12)$$

where μ_1, θ_1, r_1 and D_1 are the bond dipole moments, dipole vector angle, dipole–charge separation and effective dielectric constant for the R_1 substrate, μ_2, θ_2, r_2 and D_2 are the corresponding quantities for the R_2 substrate and the dipole–dipole term involving X_1 has been neglected, since it rarely exceeds 10 % of the charge–dipole term.

Since the Taft substituent parameters are given by

$$2{\cdot}303 RT \rho^*_{CH_2X}(\sigma^*_{R_1} - \sigma^*_{R_2}) \simeq -\partial_R \Delta E_X \qquad (3.13)$$

and the $\rho^*_{CH_2X}$ can in principle be expressed on the Hammett scale, it would seem feasible to attempt to calculate substituent parameters. There are several problems that have already been mentioned which prevent any useful progress along these lines. These may be summarized as:

(i) uncertainty as to the location of the point charge of the reaction site and the point dipole of the substituent;

(ii) the approximate nature of the point charge-point dipole treatment;

(iii) the uncertain magnitude of the group dipole moments; and

(iv) the problem of effective dielectric constant.

In the case of aromatic substrates, there are strong indications that direct electrostatic interactions are also important. This is particularly evident from studies in naphthalene derivatives described in Chapter 2 and recent results reported by Golden and Stock (1966) on the strengths of the acids

If, because of dipole orientation and dielectric constant, the direct electro-static contribution to the $8,\beta$-interaction in naphthalene is negligible and effects arising from other sources are essentially equal for $6,\beta$ and $8,\beta$-interactions, a very approximate estimate can be made that 30–40% of the $6,\beta$-effect is electrostatic in origin. Similarly from a comparison of properly scaled σ^* values for the bicyclo-octane derivatives with the corresponding σ_m and σ_p values, one concludes that about 40% of the aromatic values arise from direct electrostatic effects.

Rejecting the σ bond system as an unimportant pathway for substituent effects leaves the π electron system as the origin of the residual effect in aromatic substrates. Perhaps a characteristic feature of π electronic effects is their alternating nature evident from a comparison of σ_m and σ_p values. This alternation is most pronounced for conjugating substituents, but is also apparent for non-conjugating substituents. Both types of substituent change the charge distribution in the aromatic system, the former by delocalization and the latter by a general distorting effect.

Despite the fact that the π electronic effects appear responsible for more than half of the total aromatic substituent effect, the fall-off factors per methylene group are similar, e.g. $\rho_{ArCH_2CH_2CO_2H}/\rho_{ArCH_2CO_2H} \simeq 0.5$ and $\rho_{ArCH_2CO_2H}/\rho_{ArCO_2H} \simeq 0.5$, in aliphatic and aromatic substrates. One way in which this might arise is if the π electronic effects also reach the reaction site by an electrostatic mechanism. The fall-off factor $\rho_{ArCH_2NH_3^+}/\rho_{ArNH_3^+} = 0.36$ is smaller corresponding to the existence of an additional mechanism for interaction between the reaction site in $ArNH_2$ and the π electron system.

This hypothesis can be incorporated into a perturbation treatment of the direct interactions of equation 3.10, thus

$$I_{R_1G}^{X_2} - I_{R_2G}^{X_2} \simeq \sum_i (\partial E/\partial\alpha_i)_{GR}^{X_2}(\alpha_i^{R_1} - \alpha_i^{R_2})$$

$$+ \sum_{ij} (\partial E/\partial\beta_{ij})_{GR}^{X_2}(\beta_{ij}^{R_1} - \beta_{ij}^{R_2})$$

$$+ P_{X_2}\sum_k (q_k^{R_1} - q_k^{R_2})g_{kX_2} \tag{3.14}$$

where the α_i^R and β_{ij}^R are the Coulombic and resonance integrals appearing in the molecular orbital treatment of the energy of the RGX system, and the summations extend over all atoms i and bonds ij of the aromatic system. P_{X_2} is a polarity function depending upon the nature of X_2 and would be Nze^2 for a pole of charge ze or $Ne\mu_{X_2}$ for a dipolar group; q_k^R is the π electron density at atom k of the linking group G; and g_{kX_2} is a function of the distance between k and X_2, the effective dielectric constant and the relative orientation when X_2 is a dipole.

Neglecting all atoms and bonds of the aromatic system other than those involving the R and X groups and the atoms r and x to which they are attached, simplifies equation 3.15 to

$$I_{R_1G}^{X_2} - I_{R_2G}^{X_2} \simeq (\partial E/\partial\alpha_R)_{GR}^{X_2}(\alpha_{R_1} - \alpha_{R_2}) + (\partial E/\partial\alpha_r)_{GR}^{X_2}(\alpha_r^{R_1} - \alpha_r^{R_2})$$

$$+ (\partial E/\beta_{rR})_{GR}^{X_2}(\beta_{rR_1} - \beta_{rR_2}) + P_{X_2}\sum_k (q_k^{R_1} - q_k^{R_2})g_{kX_2} \quad (3.15)$$

A pure inductive or non-conjugating substituent will be one for which the R group is not part in the aromatic system. Thus the terms in $(\alpha_{R_1} - \alpha_{R_2})$ and $(\beta_{rR_1} - \beta_{rR_2})$ will be absent and the charge density summation in the final term will not include R. Similar considerations apply in the case of non-conjugating reaction side chains X_2 and X_1.

Setting $(\alpha_r^{R_1} - \alpha_r^{R_2}) = \epsilon(\alpha_{R_1} - \alpha_{R_2})$ and $(\alpha_x^{X_2} - \alpha_x^{X_1}) = \epsilon(\alpha_{X_2} - \alpha_{X_1})$, where ϵ is a σ bond inductive relay factor for which ca. 0·1 may be a suitable value, one can write for non-conjugating reaction side chains

Non-conjugating substituents

$$\delta_R\Delta E_X \simeq \epsilon(\alpha_{R_1} - \alpha_{R_2})[2\epsilon\pi_{r,x}(\alpha_{X_2} - \alpha_{X_1}) + P_{X_2}\sum_k \pi_{r,k}g_{kX_2} - P_{X_1}\sum_k \pi_{r,k}g_{kX_1}]$$

$$+ (P_{R_1} - P_{R_2})[\epsilon(\alpha_{X_2} - \alpha_{X_1})\sum_k \pi_{x,k}g_{kR} + P_{X_2}g_{RX_2} - P_{X_1}g_{RX_1}] \quad (3.16)$$

Conjugating substituents

$$\delta_R\Delta E_X \simeq (\alpha_{R_1} - \alpha_{R_2})[2\epsilon(\alpha_{X_2} - \alpha_{X_1})(\pi_{R,x} + \epsilon\pi_{r,x}) + P_{X_2}(\sum_k \pi_{R,k}g_{kX_2} + \epsilon\sum_k \pi_{r,k}g_{kX_2})$$

$$- P_{X_1}(\sum_k \pi_{R,k}g_{kX_1} + \epsilon\sum_k \pi_{r,k}g_{kX_1})]$$

$$+ (\beta_{rR_1} - \beta_{rR_2})[3\epsilon\pi_{rR,x}(\alpha_{X_2} - \alpha_{X_1}) + P_{X_2}\sum_k \pi_{rR,k}g_{kX_2} - P_{X_1}\sum_k \pi_{rR,k}g_{kX_1}]$$

$$+ (P_{R_1} - P_{R_2})[\epsilon(\alpha_{X_2} - \alpha_{X_1})\sum_k \pi_{x,k}g_{kR} + P_{X_2}g_{RX_2} - P_{X_1}g_{RX_1}] \quad (3.17)$$

where

$$\pi_{i,p} = (\partial^2 E/\partial\alpha_i\partial\alpha_p) = (\partial q_i/\partial\alpha_p) = \pi_{p,i}$$

are the atom–atom polarizabilities

$$\pi_{pq,i} = \tfrac{1}{2}(\partial^2 E/\partial\alpha_i\partial\beta_{pq}) = \tfrac{1}{2}(\partial q_i/\partial\beta_{pq}) = \tfrac{1}{2}\pi_{i,pq}$$

are the bond–atom polarizabilities in RGX; and

$$g_{kR} \simeq g_{kR_1} \simeq g_{kR_2}$$

$$g_{RX_2} \simeq g_{R_1X_2} \simeq g_{R_2X_2}$$

$$g_{RX_1} \simeq g_{R_1X_1} \simeq g_{R_2X_1}$$

are the appropriate functions of distance, angle and effective dielectric constant, which are assumed to be essentially equal for a typical R_1/R_2 pair.

Equation 3.17 may be summarized as

$$\partial_R \Delta E_X \simeq A_X(\alpha_{R_1} - \alpha_{R_2}) + B_X(\beta_{rR_1} - \beta_{rR_2}) + C_X(P_{R_1} - P_{R_2}) \qquad (3.18)$$

in which the factors A_X, B_X and C_X are determined by the reaction under consideration, the linking group and the relative disposition of substituent and reaction site.

For the standard reaction of the Hammett equation defining the σ parameters

$$-2 \cdot 303 R T_0(\sigma_{R_1} - \sigma_{R_2}) \simeq \partial_R \Delta E_0$$

$$\simeq A_0(\alpha_{R_1} - \alpha_{R_2}) + B_0(\beta_{rR_1} - \beta_{rR_2}) + C_0(P_{R_1} - P_{R_2})$$

$$(3.19)$$

and for the reaction series corresponding to $X_1 \to X_2$ which follows the Hammett relationship

$$\frac{T_X}{T_0}\rho_{X_2X_1} \simeq \frac{A_X(\alpha_{R_1} - \alpha_{R_2}) + B_X(\beta_{rR_1} - \beta_{rR_2}) + C_X(P_{R_1} - P_{R_2})}{A_0(\alpha_{R_1} - \alpha_{R_2}) + B_0(\beta_{rR_1} - \beta_{rR_2}) + C_0(P_{R_1} - P_{R_2})} \qquad (3.20)$$

But $\rho_{X_2X_1}$ is independent of R_1 and R_2, so that either, (i) two of the terms in equations 3.18 and 3.19 are negligible, or (ii)

$$A_X/A_0 \simeq B_X/B_0 \simeq C_X/C_0 \simeq T_X\rho_{X_2X_1}/T_0 \qquad (3.21)$$

Inspection of the quantities in equation 3.17 that determine the factors A, B and C suggest that even this approximate expression must be subject in real cases to extensive simplification. It seems possible that both conditions (i) and (ii) approximately fulfilled may be responsible for observation of Hammett correlations. The nature of these simplifications is not at present clear, and attempts to calculate substituent and reaction parameters are premature. The current problem may be summarized as a lack of information as to the correct atom–atom and bond–atom polarizabilities, difficulties in the specification of the g functions to be employed and the absence of extensive information on systems other than benzene with which to test the above model and identify the simplifications. Very limited success has so far attended the more simplified approaches to inductive effects by Peters (1957) in terms of σ bond relay of atom–atom polarizations, and to resonance effects by Roberts and Jaffé (1963) in terms of direct electrostatic interactions arising from π electron density changes. The present treatment in a sense combines these ideas.

Some justification can be obtained for attempts to separate the total substituent effective into inductive and resonance contributions. If the quantity given in equation 3.16 is symbolized $[\delta_R \Delta E_X]_I$, and regarded as an inductive component, then the total substituent effect for conjugating groups, equation 3.17, may be analysed into

$$\delta_R \Delta E_X \simeq [\delta_R \Delta E_X]_I + [\delta_R \Delta E_X]_R \qquad (3.22)$$

where

$$[\delta_R \Delta E_X]_R \simeq (\alpha_{R_1} - \alpha_{R_2})[2\epsilon\pi_{R,X}(\alpha_{X_2} - \alpha_{X_1}) + P_{X_2}\sum_k \pi_{R,k}g_{kX_2} - P_{X_1}\sum_k \pi_{R,k}g_{kX_1}]$$

$$+ (\beta_{rR_1} - \beta_{rR_2})[3\epsilon\pi_{rR,X}(\alpha_{X_2} - \alpha_1) + P_{X_2}\sum_k \pi_{rR,k}g_{kX_2}$$

$$- P_{X_1}\sum_k \pi_{rR,k}g_{kX_1}] \qquad (3.23)$$

has the characteristics of a resonance component.

Inductive parameters given by

$$(\sigma_{R_1}^I - \sigma_{R_2}^I) \simeq -[\delta_R \Delta E_X]_I / 2 \cdot 303 R T_X \rho_{X_2 X_1} \qquad (3.24)$$

might well show a close proportionality to the Taft σ^* parameter differences (cf. equation 3.13), by virtue of the common $(P_{R_1} - P_{R_2})$ factor and a rough relationship between $(P_{R_1} - P_{R_2})$ and $(\alpha_{R_1} - \alpha_{R_2})$, which to some extent will determine differences in bond dipole moments. The relationship

$$\sigma_p^I / \sigma_m^I \simeq 1 \cdot 15 \qquad (3.25)$$

suggested by Exner (1966) also appears reasonable.

Taft's σ_R parameters although obtained by assuming $\sigma_p^I = \sigma_m^I$ (see Chapter 2) may still be useful for substituents with numerically large σ_R but small σ_I values and approximate closely to resonance parameters given by

$$(\sigma_{R_1}^R - \sigma_{R_2}^R) \simeq -[\delta_R \Delta E_X]_R / 2 \cdot 303 R T_X \rho_{X_2 X_1} \qquad (3.26)$$

Equation 3.23 indicates that for sets of substituents having an essentially constant $(\beta_{rR_1} - \beta_{rR_2})$ the resonance component will be proportional to $(\alpha_{R_1} - \alpha_{R_2})$, and hence to the difference in group electronegativities (cf. Wells, 1968). Figure 3.1 indicates that Taft's σ_R values for the first-row donor groups display just this behaviour and there are indications that the second-row donors may follow a parallel relationship.

All considerations discussed above apply to the effect of substituent groups possessing distinct polar character or pronounced conjugating ability. Simple alkyl groups certainly cannot be treated in this way. These are essentially non-polar and the status of hyper-conjugation is still unclear. Several anomalous features (see Chapter 2) imply that factors neglected in the case of polar substituents need to be taken into account. One such factor may be the incomplete cancellation of steric effects particularly those affecting

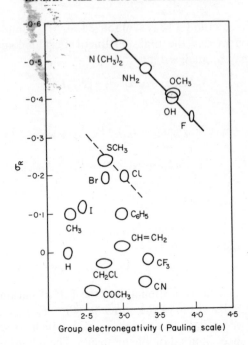

FIG. 3.1. Relationship between σ_R values and group electronegativities.

solvation. Another has been examined by Brown (1959) in terms of dispersion interactions arising from polarizability. Both factors are more difficult to assess than polar effects but appear to be important.

REFERENCES

Bell, R. P., and Wright, G. A. (1961). *Trans. Faraday Soc.*, **57**, 1377.

Brown, T. L. (1959). *J. Am. chem. Soc.*, **81**, 3229.

Exner, O. (1966). *Colln. Czech. chem. Commun.*, **31**, 65.

Golden, R., and Stock, L. M. (1966). *J. Am. chem. Soc.*, **88**, 5928.

Hepler, L. G. (1963). *J. Am. chem. Soc.*, **85**, 3089.

Holtz, H. D., and Stock, L. M. (1964). *J. Am. chem. Soc.*, **86**, 5186.

Kirkwood, J. G., and Westheimer, F. H. (1938). *J. chem. Phys.*, **6**, 506, 513.

Lewis, G. N. (1916). *J. Am. chem. Soc.*, **38**, 782.

Peters, D. (1957). *J. chem. Soc.*, 2654.

Roberts, J. L., and Jaffé, H. H. (1963). *Tetrahedron*, **19**, Supp. 2, 455.

Skinner, H. A., and Pilcher, G. (1963). *Quart. Rev.*, **17**, 264.

Waters, W. A. (1934). *J. chem. Soc.*, 1551.

Wells, P. R. (1968). *Prog. phys. org. Chem.*, **6**, in press.

4

Correlation of Changes in the Reaction Medium

By far the larger part of chemistry is that pertaining to the liquid phase, and the dominant phases continue to be solutions in water or mixtures of water and other liquids. Unfortunately the liquid phase is much less well understood than the gas and solid phases. It possesses neither the definite structure of the solid phase nor the almost random motion of the gas phase.

Many reactions that take place in solution are unknown in the gas phase, and reactions apparently common to both phases are often mechanistically quite different. There is a certain amount of common ground in the case of non-polar solvents, e.g. hydrocarbons and halogenated hydrocarbons, but even in these cases some marked differences are observed. Little can be said at present concerning reactions in the solid phase, since few studies have yet been described where local melting does not occur in the regions where reaction takes place.

The problem of understanding the influence of reaction medium on reactivity, i.e. solvent effects, resolves itself to one of recognizing the precise functions of the solvent molecules, both individually and collectively, and of devising some means of measuring these functions. Various properties of pure and mixed solvents are readily measured with high precision, e.g. composition, density, refractive index, conductivity, dielectric constant, viscosity and so on, but the relationships between these properties and the required functions are complex.

As in the case of substituent effects, one hopes that the observation of linear free-energy correlations will provide information assisting in the construction and testing of theoretical treatments.

The free-energy changes concerned in the study of solvent effects may be expressed in terms of free energies of solvation. For some reaction, represented by the conversion of reactants, R, to products, P, taking place in the gas phase or in solvents, S or S', the relative reactivities may be assessed in terms of the equilibrium constants K_g, K_S and $K_{S'}$

$$\log K_g = -\Delta F_g/2\cdot303RT$$

$$\log K_S = -\Delta F_S/2\cdot303RT$$

$$\log K_{S'} = -\Delta F_{S'}/2\cdot303RT \tag{4.1}$$

57

(As usual, relative rates would be assessed in terms of rate constants and free energies of activation, in which case P would represent a transition state configuration.)

The quantities in equation 4.1 are related by

$$\Delta F_S = \Delta F_g + \Delta F_{sol}(P;S) - \Delta F_{sol}(R;S)$$

$$\Delta F_{S'} = \Delta F_g + \Delta F_{sol}(P;S') - \Delta F_{sol}(R;S') \qquad (4.2)$$

where ΔF_g is the molar free energy change of the reaction and $\Delta F_{sol}(X;S)$ are the molar free-energy changes accompanying the transfer of 1 mole of X from the gas phase to a solution in S, i.e. the molar free energy of solution of X in S.

It follows that the difference between solution-phase and gas-phase reactivity

$$\log(K_S/K_g) = -[\Delta F_{sol}(P;S) - \Delta F_{sol}(R;S)]/2.303RT \qquad (4.3)$$

and between solvents

$$\log(K_S/K_{S'}) = -[\Delta F_{sol}(P;S) - \Delta F_{sol}(P;S')]/2.303RT$$

$$+ [\Delta F_{sol}(R;S) - \Delta F_{sol}(R;S')]/2.303RT \qquad (4.4)$$

arises from the free energies of solution of the species concerned.

This analysis provides no answers to the solvent-effect question. It simply changes the form of the question to one of accounting for the phenomenon of solvation. The factors believed to be involved range from the reduction of the electrostatic free energy of ionic species by dielectric effects to van der Waal's attractive forces and covalent bonding of the donor–acceptor type. Particularly important in the latter category are hydrogen-bonding interactions and, in the case of reaction rate processes, the direct involvement of the solvent as a reagent in a rate-controlling step.

An alternative expression can be written in terms of activity coefficients (f) relative to some standard conditions of solvent and concentration, i.e. K_0, so that

$$\log(K_S/K_0) = \log(f_R^S/f_P^S) \qquad (4.5)$$

and

$$\log(K_S/K_{S'}) = \log(f_R^S/f_P^S) - \log(f_R^{S'}/f_P^{S'}) \qquad (4.6)$$

Again the question of solvent effects has simply been rephrased, this time in terms of activity coefficients, i.e. relative deviations from some selected ideal situation.

The simplest model for solvent effects on ionic reactions is the electrostatic treatment in which the solvent is treated as a continuum of dielectric constant,

D. The electrostatic free-energy change in transferring 1 mole of ions (X) of charge ze and radius r to this medium from the gas phase $(D = 1)$ is employed as an approximation for the molar free energy of solution, thus

$$\Delta F_{sol}(X) \simeq -\frac{Nz^2e^2}{r}(1 - 1/D) \qquad (4.7)$$

Assuming this to be the most important factor, equation 4.4 becomes

$$\log(K_S/K_{S'}) = \frac{Ne^2}{2 \cdot 3RT}\left[\frac{z_P^2}{r_P} - \frac{z_R^2}{r_R}\right]\left(\frac{1}{D'} - \frac{1}{D}\right) \qquad (4.8)$$

This corresponds to $K_S > K_{S'}$ for a reaction producing ions, i.e. $z_P^2 > z_R^2$, whenever $D > D'$.

An expression similar to, but more complex than, equation 4.7 can be written for the electrostatic free energy of a dipolar molecule, and essentially the same conclusions can be reached concerning the influence of dielectric constant.

The macroscopic properties of the solvent, e.g. its dielectric constant, are frequently inappropriate at the intermolecular level where solvent–solute interactions take place. In addition, the composition of a mixed solvent in the immediate vicinity of a solute molecule may be quite different from that of the overall solvent. This can arise at the two extremes from weak attractive forces or from complex formation involving the solute and one component of the solvent in preference to the other. For these reasons it is therefore likely that simple expressions, such as equation 4.8, will at best give only a rough account of solvent effects.

4.1 THE CORRELATION OF SOLVOLYSIS RATES: GRUNWALD–WINSTEIN EQUATION

For most purposes the mechanisms for nucleophilic substitution at saturated carbon atoms can be divided into two types labelled S_N1 and S_N2 (cf. Hine, 1962; Gould, 1960). In the S_N1 mechanism, it is believed that the rate-controlling step is the ionization of the substrate to form a carbonium ion with expulsion of the leaving group. This process may lead to ion pairs or separated ions whose subsequent reactions to yield the observed products are rapid. In the S_N2 mechanism, entry of the reagent and expulsion of the leaving group are concurrent. Generally the kinetic dependence on reagent concentration serves to differentiate between the two possibilities. In solvolyses, however, there will be no perceptible change in solvent composition and hence a kinetic order with respect to solvent cannot be determined. Several other diagnostic tests of mechanism have been employed, such as stereochemical result, substituent effects, salt effects and the tendency towards rearrangement. These are often inapplicable or ambiguous, and it has been suggested that solvent effects on reactivity may be of assistance.

The basis for this suggestion is that reactions leading to the formation of ions or to substantial local increases in charge density should be sensitive to the polarity and ion solvating power of the solvent. Further, the ionization process of the S_N1 reaction should be more sensitive to these properties than the concerted S_N2 solvolysis. In the latter case the change in ionic character of the substrate is probably less severe and other solvent properties, i.e. nucleophilicity, are involved.

The rate of solvolysis (k_s^0) of t-butyl chloride in a particular solvent relative to the rate (k_0^0) in 80% v/v aqueous ethanol was chosen by Grunwald and Winstein (1948) as a measure of "solvent ionizing power" and given the symbol Y. Thus

$$Y_s \equiv \log(k_s^0/k_0^0) \tag{4.9}$$

and the Y values listed in Tables 4.1 and 4.3 were obtained in this way by Fainberg and Winstein (1956). The reference solvent, $Y = 0$, is 80% v/v aqueous ethanol, i.e. 80 volumes of ethanol and 20 volumes of water, which is a commonly used solvent in studies of other substrates, and lies about half way in the range between the best, water, and poorest solvents in terms of solvolysis rate.

From the nature of its definition, Y will have positive values where solvolysis is more rapid than in the reference solvent. Increased solvolysis rate is associated with increased ion solvating power for the S_N1 mechanism.

TABLE 4.1. SOLVENT PARAMETERS (AQUEOUS SOLVENTS)
$Y \equiv \log(k_s^0/k_0^0)$ for $(CH_3)_3CCl$ at 25°C

% v/v of added solvent	Added solvent					
	Ethanol	Methanol	Dioxan	Acetone	Acetic Acid†	Formic Acid†
0	3·49	3·49	3·49	3·49	3·49	3·49
10	3·31	3·28	3·22	3·23	..	..
20	3·05	3·03	2·88	2·91	..	..
25	2·91	..	..	2·69	2·84	3·10
50	1·66	1·97	1·36	1·40	1·94	2·64
60	1·12	1·49	0·72	0·80	1·52	..
80	0·00‡	0·38	−0·83	−0·67	..	2·32
90	−0·75	−0·30	−2·03	−1·86	..	2·22
95	−1·29	..	..	−2·76	..	2·16
100	−2·03	−1·09	..	..	−1·64	2·05

† Containing ca. 0·006M lithium salt. ‡ By definition.

If the **Y** values do indeed measure a solvent property and are independent of the substrate undergoing solvolysis, then, provided the S_N1 mechanism prevails, one would expect for some other substrate (RX) that

$$\log(k_s/k_0) = m_{RX}Y_s \qquad (4.10)$$

In equation 4.10 m_{RX} is a proportionality factor that depends upon the substrate. From the first, however, it was observed that a unique m parameter for each substrate applicable to all solvents was not obtained. Instead the correlation of log k with **Y** generally yields a series of almost parallel lines, one for each binary solvent pair. This corresponds to small variations in m and often larger variations in the intercept, log k_0, i.e. the predicted rate of solvolysis in the standard solvent. These variations are illustrated in Table 4.2 (cf. Winstein *et al.*, 1957).

It is convenient first to examine the **Y** values and subsequently the performance of equation 4.10.

The **Y** values for pure solvents are listed in Table 4.3 together with the dielectric constants, the enthalpy and entropy of activation for the solvolysis of t-butyl chloride in these solvents reported by Winstein and Fainberg (1957).

The unique character of water observed in many processes is evident once again. It is a substantially better solvent for the solvolysis reaction than any other, owing to the large positive entropy of activation. The other pure solvents fall into two groups, the protic solvents, i.e. alcohol and acids, and the aprotic solvents. The results for formamide require that it be placed in the first group.

TABLE 4.2. SUBSTRATE PARAMETERS (25°C)

	Aq. ethanol m	Aq. ethanol $\log k_0$	Aq. dioxan m	Aq. dioxan $\log k_0$	Aq. acetic acid m	Aq. acetic acid $\log k_0$
1. $(CH_3)_3CCl$	1·00	−5·03	1·00	−5·03	1·00	−5·03
2. $(CH_3)_3CBr$	0·94	−3·46	0·92	−3·34	1·07	−3·78
3. $C_6H_5(CH_3)CHCl$	0·97	−4·94	1·14	−5·49	1·14	−5·18
4. $C_6H_5(CH_3)CHBr$	0·82	−3·64	1·01	−4·14	1·25	−4·36
5. $C_6H_5(CH_3)_2CCH_2Cl$†	0·83	−8·54	0·96	−8·97	0·73	−8·11
6. $C_6H_5(CH_3)_2CCH_2Br$†	0·81	−6·97	0·93	−7·37	0·83	−6·87
7. $(C_6H_5)_2CHCl$	0·74	−2·76	1·05	−3·67	1·56	−2·66
8. $(C_6H_5)_2CHBr$	..	..	0·95	−2·44	1·69	−2·02
9. $(C_6H_5)_2CHF$	0·98	−6·56	..	..	..	..
10. $(C_6H_5)_3CF$	0·89	−3·53	1·58‡	−5·02‡	..	..
11. $(C_6H_5)_3COCOCH_3$	0·50	−3·28	0·83‡	−4·28‡	..	..

† At 50°C. ‡ Aqueous acetone.

TABLE 4.3. SOLVENT PARAMETERS (PURE SOLVENTS)

Solvent	Y	$D(25°C)$	$\Delta H^{\ddagger}$ (kcal. mole^{-1})	$\Delta S^{\ddagger}$ (e.u)
H_2O	3·49	78·5	23·2	+12·2
HCO_2H	2·05[a]	58[e]	21·0	−1·7
$C_2F_7CO_2H$	1·7	8[f]	..	..
$HCONH_2$	0·6	109[g]	22·4	−3·8
CH_3OH	−1·09	33	24·9	−3·1
CH_3CO_2H	−1·7[b]	6[g]	25·8	−2·5
C_2H_5OH	−2·03	24	26·1	−3·2
$i\text{-}C_3H_7OH$	−2·7	18	..	..
$t\text{-}C_4H_9OH$	−3·3	11[h]	..	..
$C_2H_5CONHCH_3$†	ca. −3[c]	164[h]	26[c]	..
$H.CON(CH_3)_2$†	−3·5	37	25	..
CH_3COCH_3	−5·2[d]	21	ca. 22[d]	$< -20^d$
Dioxan	−5·8[d]	2	ca. 22[d]	$< -20^d$

a. Containing 0·006M lithium salt. b. Estimated for zero salt concentration. c. Estimated from t-butyl bromide solvolysis. d. Extrapolation of aqueous solvent results. e. At 16°C. f. CF_3CO_2H value. g. At 20°C. h. At 30°C.
† Ross and Labes (1957).

The aprotic solvents yield low solvolysis rates, i.e. Y large and negative, which appear to be largely associated with a negative entropy of activation. Perhaps in this group of solvents the difference in Y can be accounted for by dielectric constant.

The protic solvents have essentially the same entropy of activation, $-(2·9 \pm 0·6)$ e.u., and the differences in Y appear to arise from differences in $\Delta H^{\ddagger}$, although in the case of formic acid the entropy of activation is distinctly less negative. Two factors appear to be involved here—dielectric constant and acidity. The former will determine the general solvating power and the latter specific interaction with the leaving group. Apparently one of the amino-hydrogens of formamide can function in the acidic sense. The other, like the single N—H of N-methylpropionamide, is believed to be tied-up in the hydrogen-bonded polymeric structure of these liquids. The difference in Y value between the alcohols is deceptive. This disappears when allowance is made for molecular weight. The value of Y/mol. wt. is essentially the same for all four in accordance with their similar acidities and dielectric constants.

For mixed solvents, no simple relationship can be found between composition and Y. Although there are linear relationships between Y and mole fraction of water (N_w) over limited composition ranges for the binary solvents, these correspond to short sections near inflexions in more complex curves.

The various aqueous solvents can be fitted to power series

$$\mathbf{Y} = a + bN_w + cN_w^2 + dN_w^3 + \dots \qquad (4.11)$$

which may be used to calculate new $\mathbf{Y}$ values. An essentially equivalent procedure is the interpolation of $\mathbf{Y}$ values from the $\mathbf{Y}$ versus composition curves. Equation 4.11 has been used to extrapolate the values of $\mathbf{Y}$ for pure acetone and dioxan given in Table 4.3.

In the water-poor solvents, the first two terms of equation 4.11 are sufficient to give a rough idea of the effect of small additions of water to the pure solvents. Thus

S	Y (approx.)
Formic acid	$2 \cdot 07 N_s + 3 \cdot 06 N_w$
Methanol	$-1 \cdot 08 N_s + 2 \cdot 28 N_w$
Ethanol	$-2 \cdot 04 N_s + 4 \cdot 05 N_w$
Acetic acid	$-1 \cdot 63 N_s + 6 \cdot 82 N_w$
Acetone	$-5 \cdot 2 N_s + 9 \cdot 86 N_w$
Dioxan	$-5 \cdot 8 N_s + 10 \cdot 2 N_w$

In view of the suggested factors contributing to the $\mathbf{Y}$ values of pure solvents it is not surprising that no correlation is observed for mixed solvents between the $\mathbf{Y}$ values and dielectric constant, D, other than a general fall in $\mathbf{Y}$ as D decreases.

Although a plot of $\mathbf{Y}$, and hence free energy of activation, $\Delta F^{\ddagger}$, versus mole fraction of water tends to be a smooth curve, this does not carry over to plots of $\Delta H^{\ddagger}$ nor $\Delta S^{\ddagger}$ versus mole fraction. Figures 4.1(a), (b) and (c) illustrate this behaviour. Over most of the composition range the increase in $\Delta F^{\ddagger}$ is made up of an increase in $-T\Delta S^{\ddagger}$ with either an increase, a decrease or a relatively steady $\Delta H^{\ddagger}$. There are however short ranges in which $-T\Delta S^{\ddagger}$ decreases while $\Delta H^{\ddagger}$ increases rapidly. For no binary solvent is a linear relationship between $\Delta H^{\ddagger}$ and $\Delta S^{\ddagger}$ observed over a substantial composition range. The behaviour for aqueous ethanol, illustrated in Fig. 4.1(d), is not an extreme case.

Following equation 4.4, the $\mathbf{Y}$ values may be expressed in terms of relative free energies of solvation, i.e.

$$2 \cdot 303 RT(\mathbf{Y}_s - \mathbf{Y}_{s'}) = -[\Delta F_{sol}(\text{BuCl}^{\ddagger}; S) - \Delta F_{sol}(\text{BuCl}^{\ddagger}; S')$$
$$+ [\Delta F_{sol}(\text{BuCl}; S) - \Delta F_{sol}(\text{BuCl}; S')] \qquad (4.12)$$

or, alternatively, in terms of activity coefficients relative to 80 % v/v aqueous ethanol as the standard state, i.e.

$$\mathbf{Y}_s - \mathbf{Y}_{s'} = \log(f_{\text{BuCl}}/f_{\ddagger})_s - \log(f_{\text{BuCl}}/f_{\ddagger})_{s'} \qquad (4.13)$$

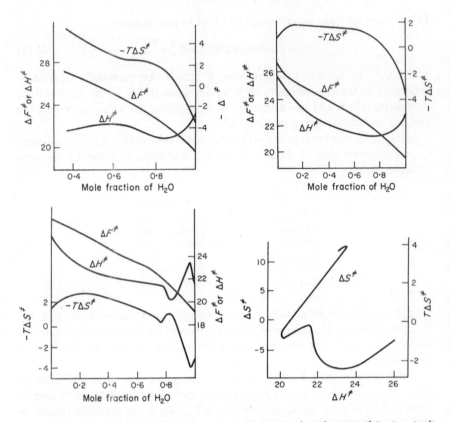

Fig. 4.1. Solvolyses of t-butyl chloride showing variation of $\Delta F^{\ddagger}$, $\Delta H^{\ddagger}$ and $T\Delta S^{\ddagger}$ (kcal. mole^{-1}) with solvent composition: (a), aqueous dioxan; (b), aqueous acetic acid; (c), aqueous ethanol; (d), aqueous ethanol.

The initial argument that the **Y** values measure ion-solvating power would lead one to expect that its dominant terms in equation 4.12 and 4.13 were those pertaining to the transition state, i.e. $\Delta F_{sol}(\text{BuCl}^{\ddagger}; S)$ and $f_{\ddagger}$. However in the comparison of solvents one is concerned with *differences* in free energies of solvation and not with their absolute values. The relative free energies of solvation of the substrates have been determined from vapour-pressure measurements, so that the quantities listed in Table 4.4 can be compared (cf. Winstein and Fainberg, 1957).

The unique behaviour of water is again evident, in that the major change from 80% aqueous ethanol to pure water takes place in the free energy of solvation of the neutral substrate molecule. The differences between the other pure solvents and between each of them and the standard is found largely in the free energy of solvation of the transition state. In the aqueous

TABLE 4.4. COMPONENTS OF RELATIVE FREE ENERGY OF ACTIVATION

$$\delta\Delta F^{\ddagger} = -2{\cdot}303RT \; Y(\text{kcal. mole}^{-1})$$

$$\delta\Delta F_{sol}(\text{BuCl}) = [\Delta F_{sol}(\text{BuCl}; S) - \Delta F_{sol}(\text{BuCl}; S')]$$

$$\therefore \quad \delta\Delta F^{\ddagger}_{sol} = \delta\Delta F^{\ddagger} - \delta\Delta F_{sol}(\text{BuCl}) = [\Delta F_{sol}(\text{BuCl}^{\ddagger}; S) - \Delta F_{sol}(\text{BuCl}^{\ddagger}; S')]$$

$$(S' \equiv 80\% \text{ aq. } C_2H_5OH \text{ at } 25°C)$$

Solvent (S)	$\delta\Delta F^{\ddagger}$	$\delta\Delta F_{sol}(\text{BuCl})$	$\therefore \quad \delta\Delta F^{\ddagger}_{sol}$
H_2O	$-4{\cdot}8$	$-3{\cdot}55$	$-1{\cdot}25$
CH_3OH	$1{\cdot}5$	$0{\cdot}6$	$0{\cdot}9$
CH_3CO_2H	$2{\cdot}25$	$0{\cdot}75$	$1{\cdot}5$
C_2H_5OH	$2{\cdot}8$	$0{\cdot}8$	$2{\cdot}0$
90% aq. C_2H_5OH	$1{\cdot}0$	$0{\cdot}65$	$0{\cdot}35$
80% aq. C_2H_5OH	$0{\cdot}0$	$0{\cdot}0$	$0{\cdot}0$
70% aq. C_2H_5OH	$-0{\cdot}8$	$-0{\cdot}45$	$-0{\cdot}35$
60% aq. C_2H_5OH	$-1{\cdot}5$	$-1{\cdot}15$	$-0{\cdot}35$
50% aq. C_2H_5OH	$-2{\cdot}25$	$-1{\cdot}75$	$-0{\cdot}5$
40% aq. C_2H_5OH	$-3{\cdot}1$	$-2{\cdot}4$	$-0{\cdot}7$
90% aq. CH_3OH	$0{\cdot}4$	$0{\cdot}2$	$0{\cdot}2$
80% aq. CH_3OH	$-0{\cdot}5$	$-0{\cdot}2$	$-0{\cdot}3$
70% aq. CH_3OH	$-1{\cdot}3$	$-0{\cdot}6$	$-0{\cdot}7$

methanol solvents, $\delta\Delta F^{\ddagger}$ is composed of essentially equal contributions from $\delta\Delta F_{sol}(\text{BuCl})$ and $\delta\Delta F^{\ddagger}_{sol}$, whereas the differences between the aqueous ethanol solvents are again largely in $\delta\Delta F_{sol}(\text{BuCl})$.

One of the most outstanding features of Fig. 4.1 is the change in $\Delta H^{\ddagger}$ in the region 0·8–1·0 mole fraction water in ethanol. Since $\Delta F^{\ddagger}$ smoothly increases through this range, there is an almost mirror image change in $-T\Delta S^{\ddagger}$. Arnett et al. (1963) has investigated the minimum observed at ca. 0·9 mole fraction by measuring heats of solution of several species in this region. For many neutral molecules, including t-butyl chloride, the relative heat of solution passes through an endothermic maximum in this region. These maxima are higher the larger the volume of the solute, and in comparison almost no maximum is observed in the relative heat of solution of an electrolyte such as $(CH_3)_4N^+Cl^-$. If the observed $\Delta H^{\ddagger}$ values are corrected for changes in the heat of solution of t-butyl chloride, the resultant relative heat of solution of the solvolysis transition state varies with solvent composition essentially as expected for "trimethylcarbonium chloride".

Equation 4.10 can only be successfully applied to other substrates if separate m values and log k_0 values are used for each binary solvent. Some of these values for S_N1 solvolyses are illustrated in Table 4.2. When log k is plotted against the Y values, not only is there dispersion that can be accommodated by a series of lines, but these lines all show a slight curvature. Plots of this type also suggest, as do the data in Table 4.2, that there is as much divergence between t-butyl chloride and bromide (substrates 1 and 2) as between t-butyl chloride and α-phenylethyl chloride (3).

The effect of the leaving group can be isolated from that of the alkyl group by the comparison

$$\log k_{RBr} = a \log k_{RCl} + b \qquad (4.14)$$

All the non-acidic solvent data fit a single line corresponding to equation 4.14, whereas the aqueous acetic acid data fit a second line; these results are summarized in Table 4.5.

TABLE 4.5. CORRELATION BY EQUATION 4.5

R-group	Temperature	Aq. alcohols and dioxan		Aq. acetic acid	
		a	b	a	b
t-butyl	0°C	0·94	1·32	1·07	1·79
α-Phenylethyl	25°C	0·93	0·96	1·08	1·19
Benzhydryl	25°C	0·94	1·06	1·09	0·88
Neophyl	50°C	0·98	1·37	1·12	2·24

In all cases a difference between non-acid and acidic solvents of equal **Y** value is found in the sense that (k_{RBr}/k_{RCl}) is lower in the acid-containing solvents. It has already been noted in connection with the **Y** values of pure solvents that acidity seems to be an important factor. The present observations can be accounted for in the same way since the expulsion of the leaving group will be assisted by hydrogen-bond formation with the solvent. The acidic solvents will tend to form the stronger hydrogen bonds and, since the chloride will gain more from this type of solvation, its solvolysis rate will be enhanced relative to that of the bromide. Fluoride is a rather poor leaving group, but is even more susceptible to hydrogen-bond assistance. The solvolysis rate of benzhydryl fluoride is 10^5 times faster in acetic acid than in ethanol, whereas benzhydryl chloride solvolyses ten times more slowly in acetic acid than in ethanol.

Two other reasons for the failure of equation 4.10 seem less important than specific leaving-group effects, but should nevertheless be borne in mind. One is that the solvolysis of t-butyl chloride could be a poor model for the S_N1 reaction if the nucleophilicity of the solvent were also important. This factor alone cannot account for the observed results, since one would need to postulate greater nucleophilic participation in the case of bromides than chlorides, which seems unlikely.

A second factor is "internal return." It has been assumed that the S_N1 solvolysis rate constant (k) corresponds to the rate constant for ionization (k_i), but if a fraction, F, of the ions formed revert to substrate then k is only $(1-F)k_i$. F will presumably depend upon both substrate structure and solvent. For neophyl substrates (5 and 6 in Table 4.2), this factor is unimportant because any return would yield the rearranged tertiary substrate that would immediately solvolyse under the reaction conditions employed. Since the neophyl substrates fit the general pattern of results (cf. Table 4.4), it seems unlikely that internal return in other cases is responsible for deviations from equations 4.10.

The greatest deviation from equations 4.10 are observed with trityl substrates (10 and 11 of Table 4.2). Markedly different m and $\log k_0$ values are required for different solvent mixtures and the correlations are in general not good. Clearly $\log(f_{BuCl}/f_{BuCl}^{\ddagger})$ gives a very poor account of the changes in $\log(f_{\phi_3CX}/f_{\phi_3CX}^{\ddagger})$. This is not surprising, since the trityl substrates are extreme examples of the S_N1 mechanism. Again the fluoride-leaving group is very susceptible to the acidic solvents, solvolysis proceeding ca. 10^3 times faster in acetic acid than in an aqueous ethanol solvent of equal Y value.

In view of the suggestion that at least two solvent properties are measured by the Y values it is reasonable that the solvolysis data could be fitted by an equation of its type

$$\log k_{RX} = m_G Y_G + m_H Y_H + \log k_{RX}^0 \qquad (4.15)$$

where Y_G measures the "general ionizing power" and Y_H and "hydrogen-bonding power" of the solvent. Each substrate would then be characterized by m_G and m_H values according to its response to these solvent properties. This approach is closely related to the correlations suggested by Swain which are discussed below.

Insofar as the Y value is a measure of the polarity of the solvent, some correlation of S_N2 solvolysis and other reactions might be expected even though solvent nucleophilicity is also important. In such cases an equation of the form

$$\log k_{RX} = mY + lN + \log k_{RX}^0 \qquad (4.16)$$

where N measures solvent nucleophilicity and l the sensitivity of the substrate of this property, might be suitable. For limited solvent variations it may arise that N/Y is essentially constant for a given substrate, in which case

$$\log k_{RX} = (m + lN/Y)Y + \log k_{RX}^0$$
$$= m'Y + \log k_{RX}^0 \qquad (4.17)$$

Equation 4.17 may account in part for the correlations of S_N2 solvolysis some of which are listed in Table 4.6.

TABLE 4.6. CORRELATION OF S_N2 SOLVOLYSIS (EQUATION 4.17)

	Aq. ethanol		Aq. acetone		
	m	$\log k_0$	m	$\log k_0$	(k_{ROH}/k_{HOAc})†
$p\text{-}O_2NC_6H_4COCl$	0·33	..	..	..	13,000
$p\text{-}O_2NC_6H_4COF$	0·53	..	..	..	10,000
CH_3SO_2Cl	..	..	0·47[d]	0·47[d]	..
CH_3Br[a]	0·26	..	..	..	300
C_2H_5Br[b]	..	..	0·34	−6·79	80
$i\text{-}C_3H_7Br$[a]	0·58	..	..	..	40
$CH_3OSO_2C_6H_5$[a]	0·22	−4·50	..	..	..
$C_2H_5OSO_2C_6H_5$[a]	0·28	−4·82	0·44[d]	−5·33[d]	..
$C_2H_5OSO_2C_7H_7$[a]	0·25	..	..	..	70
$C_6H_5CH_2OSO_2C_7H_7$	0·39	..	0·65	−4·71	30
$(CH_3)_2S^+OH^-$[c]	−0·78	−1·73	..	..	..
$(C_2H_5)_2S^+OH^-$[c]	−0·84	−1·71	..	..	..

a. At 50°C; b. At 40°C; c. At 100°C; d. In aq. dioxan.
† Ratio of solvolysis rates in aq. ethanol and acetic acid of equal Y.

The outstanding features of Table 4.6 are the small m values corresponding to a reduced sensitivity to solvent polarity, the high values of k_{ROH}/k_{HOAc} arising from the low nucleophilicity of acetic acid relative to ethanol and the negative m values for the sulphonium cation decompositions where ionic character is reduced in the transition state.

4.2 THE SWAIN–MOSELY–BOWN EQUATION

In view of the discussion above, an all-embracing equation for solvent effects on solvolysis should contain parameters describing general solvent polarity, hydrogen-bonding capacity or electrophilicity, and nucleophilicity. This would require six parameters and could be derived from equations 4.15 and 4.16 in the form

$$\log k_{RX} = m_G Y_G + m_H Y_H + lN + \log k_{RX}^0 \qquad (4.18)$$

The difficulty with multi-parameter equations, such as 4.18, is that each solvent parameter tends to be set up to fit an extreme case with the other parameters simply serving as correcting factors for interactions not quite the same as this extreme. Excellent correlations are readily achieved, but the derived parameters are often of little interpretive value, and may sometimes lead to absurd conclusions.

Swain *et al.* (1955) did not go as far as equation 4.18, but suggested a four-parameter equation

$$\log k_{RX} = c_1 d_1 + c_2 d_2 + \log k_{RX}^0 \tag{4.19}$$

in which c_1 and c_2 are substrate parameters and d_1 and d_2 are solvent parameters, supposedly measuring nucleophilicity and electrophilicity, respectively. Again the standard solvent 80% aqueous ethanol was chosen, i.e. $\log k_{RX}^0$, and the scales of the various parameters were set by putting $c_1 = 3 \cdot 00 c_2$ for CH_3Br, $c_1 = c_2 = 1 \cdot 00$ for t-C_4H_9Cl and $3 \cdot 00 c_1 = c_2$ for $(C_6H_5)_3CF$. The reasoning behind these assignments was that a large part of the failure of the Grunwald–Winstein equation arises from classification of t-butyl chloride in the "limiting" S_N1 category. Its behaviour relative to the limiting S_N2, represented by CH_3Br, and to trityl fluoride was taken to imply that its solvolysis mechanism could be regarded as intermediate between the limiting S_N2 and S_N1 types, the latter being better represented by trityl fluoride. Table 4.7 illustrates some of the parameters obtained.

From the manner in which the parameter scales were set up, it was intended that d_1 and d_2 should be rough measures of solvent nucleophilicity and electrophilicity, respectively, and that c_1 and c_2 would measure the sensitivity

TABLE 4.7. SWAIN–MOSELY–BOWN PARAMETERS

Substrate	c_1	c_2	Solvent	d_1	d_2
$(C_6H_5)_3CF$	0·37[a]	1·12[a]	H_2O	−0·44	4·01
$(C_6H_5)_3COAr$[b]	0·18	0·59	40% C_2H_5OH	−0·26	2·13
$(C_6H_5)_3COCOCH_3$	2·19	0·77	50% C_2H_5OH	+0·12	1·33
$(C_6H_5)_2CHF$	0·32	1·17	60% C_2H_5OH	−0·22	1·34
$(C_6H_5)_2CHCl$	1·24	1·25	80% C_2H_5OH	0·00[f]	0·00[f]
$C_6H_5(CH_3)CHCl$	1·47	1·75	90% C_2H_5OH	−0·01	−0·54
$(CH_3)_3CCl$	1·00[c]	1·00[c]	C_2H_5OH	−0·53	−1·03
$C_6H_5CH_2OTos$[d]	0·69	0·39	70% CH_3OH	−0·06	−1·32
C_2H_5OTos[d]	0·65	0·24	97% CH_3OH	−0·11	−0·05
$C_6H_5CH_2Cl$	0·74	0·44	CH_3OH	−0·05	−0·73
i-C_3H_7Br	0·90	0·58	50% CH_3COCH_3	−0·25	0·97
n-C_4H_9Br	0·77	0·34	70% CH_3COCH_3	−0·09	−0·75
C_2H_5Br	0·80	0·36	80% CH_3COCH_3	−0·45	−0·68
CH_3Br	0·80[e]	0·37[e]	90% CH_3COCH_3	−0·53	−1·52
C_6H_5COF	1·36	0·66	CH_3CO_2H	−4·82	3·12
C_6H_5COCl	0·81	0·52	HCO_2H	−4·40	6·53
$ArCOCl$[b]	1·67	0·49	83% HCO_2H	−4·44	6·26
$ArCOF$[b]	1·09	0·21	..	..	..

a. $c_2/c_1 = 3$ by definition. b. $Ar \equiv C_6H_4\text{-}p\text{-}NO_2$. c. c = 1 by definition. d. Tos = *p*-toluene-sulphonate. e. $c_1/c_2 = 3$ by definition. f. $d_1 = d_2 = 0$ by definition.

to these properties. The general solvent polarity is presumably largely included in d_2.

For most substrates, in all except the acidic solvents, the c_2d_2 term in equation 4.19 is much larger than the c_1d_1 term. So much so that correlations employing the c_2d_2 term alone are only slightly inferior to the two-term correlations. The results are essentially the same as correlations with the Y values. The c_1d_1 term and the small differences between c_2d_2 and mY serve as correcting terms to improve correlations between $\log(k/k_0)$ and Y. The large value of d_2 for acidic solvents accounts for the high relative solvolysis rates of trityl and benzhydryl fluoride in these solvents, and the low values of d_1 bring $d_1 + d_2$ back roughly in accord with that expected for a polar solvent with little nucleophilic power.

If one takes the alternative view that its differences between t-butyl chloride and trityl fluoride arises not from the non-limiting S_N1 nature of the former solvolysis but from specific leaving-group effects in the latter, then no mechanistic conclusions can be drawn from the values of c_1 and c_2. The values obtained are certainly inconsistent with other mechanistic ideas. For example, the c_1 values suggest that t-butyl, benzhydryl and, most alarmingly, the bridgehead 1-bromobicyclo[2.2.2]octane substrates are most sensitive to nucleophilic attack than the typical S_N2 substrates.

As far as the correlation of data is concerned, equation 4.19 is an improvement on the simple mY treatment. If applied solely to S_N1 solvolysis, it would perform essentially the same function as the variable m correlations. When applied to both S_N1 and S_N2, it has some predictive value, but is otherwise not particularly useful. The substrate changes, solvent changes and, above all, the mechanistic changes are probably far too severe for a single LFER to apply to all data.

4.3 SOLVENT EFFECTS ON RELATIVE ACID STRENGTH

The inherent assumption in the mY treatment of solvolysis reactions is that the logarithms of activity-coefficient ratios, such as those in equation 4.5, can be expressed as the product of factors characteristic of the solvent and of the substrate (*activity postulate*). In the ionization of an acid, HA, equation 4.5 may be written as

$$pK_s - pK_0 = \log f_H + \log(f_{A^-}/f_{HA}) \tag{4.20}$$

where K_s and K_0 are the ionization constants for infinitely dilute solutions in the solvent S and the standard solvent, e.g. water, respectively. These activity coefficients are not those customarily used to account for concentration effects, but, as with all activity coefficients, they are used to account for deviations from some chosen behaviour.

The activity postulate for a series of similar acids, HA, HA', etc., may be expressed as

$$\frac{1}{m_A} \cdot \log\left(\frac{f_{A^-}}{f_{HA}}\right) = \frac{1}{m_{A'}} \cdot \log\left(\frac{f_{A'^-}}{f_{HA'}}\right) \equiv Y_- \tag{4.21}$$

where m_A depends upon the substrate and Y_- upon the solvent. The subscript of the solvent parameters, Y_-, serves to indicate that in the type of acid ionization correlated the conjugate base is negatively charged.

Combination of equations 4.20 and 4.21 yields

$$pK_s - pK_0 = \log f_H + m_A Y_- \tag{4.22}$$

and for the comparison of two acids HA and HA'

$$(pK_s - pK_0)_A - (pK_s - pK_0)_{A'} = (m_A - m_{A'})Y_- \tag{4.23}$$

In a similar manner the solvent effect upon the ionization of charged acids, BH^+, can be expressed by

$$pK_s - pK_0 = \log f_H + m_B Y_0 \tag{4.24}$$

$$\frac{1}{m_B} \cdot \log\left(\frac{f_B}{f_{BH^+}}\right) = \frac{1}{m_{B'}} \cdot \log\left(\frac{f_{B'}}{f_{B'H^+}}\right) \equiv Y_0 \tag{4.25}$$

$$(pK_s - pK_0)_B - (pK_s - pK_0)_{B'} = (m_B - m_{B'})Y_0 \tag{4.26}$$

Grunwald and Berkowitz (1951) examined the utility of this approach through equation 4.23 and found that the observed data for aqueous ethanol could be fitted with a mean probable error of fit of $0.02\, pK$ units, which is as good as the estimated experimental error in the pK differences. A similar investigation of equation 4.26 covering substituted ammonium and anilinium cations in aqueous ethanol by Gutbezahl and Grunwald (1953) gave essentially the same results.

The above correlations were carried out without absolute m values. Instead $(m_A - m_{A_0})$ and $(m_B - m_{B_0})$ values were obtained, where HA_0 and B_0 are acetic acid and aniline, respectively. The solvent parameter scales were set with $Y_- = Y_0 = 0.00$ for pure water, $Y_- = 1.00$ and $Y_0 = -1.00$ for pure ethanol.

Presumably the Y_- and Y_0- values were given opposite signs, since for the former the ionic factor f_{A^-} is in the numerator of equation 4.21, whereas for the latter the ionic factor is f_{BH^+} in the denominator of equation 4.25. For all solvents the $(pK_s - pK_0)_{HA}$ values are positive, whereas for all except pure ethanol the $(pK_s - pK_0)_B$ values are negative.

Subsequently Gutbezahl and Grunwald combined equations 4.20 and 4.24, such that for each solvent, S

$$(pK_s - pK_0)_{HA} - (pK_s - pK_0)_B = m_A Y_- m_B Y_0 \tag{4.27}$$

where m_A and m_B were determined by multiple-regression methods. Since equation 4.27 fits the experimental data as well as equations 4.23 and 4.26 it is believed that individual m values can be determined with some confidence. This makes possible the calculation of f_{H^+} values for the various aqueous ethanol solvents.

Table 4.8 lists the Y_- and Y_0 values, and Table 4.9 gives the m values for the acids correlated by means of equations 4.23 and 4.26.

TABLE 4.8. SOLVENT PARAMETERS FOR AQUEOUS ETHANOL

% w/w ethanol	Y_-	ΔpK_0^0†	Y_0	ΔpK_0^0‡	$\log f_{H^+}$
0	0·00§	0·00	0·00§	0·00	0·00†
20	0·35	0·37	−0·06	−0·22	0·01
35	0·59₅	0·67	−0·13₅	−0·48	0·04
50	0·81₅	1·08	−0·26₅	−0·72	0·25
65	0·92₅	1·53	−0·38	−0·84	0·54
80	0·96₅	2·11	−0·57	−0·89	1·15
95	..	..	−0·88	−0·45	2·9
100	1·00§	5·56	−1·00§	+1·06	4·71

† $\Delta pK_0^0 = pK_{EtOH} - pK_{H_2O}$ for acetic acid. ‡ $\Delta pK_0^0 = pK_{EtOH} - pK_{H_2O}$ for anilinium cation.
§ By definition.

TABLE 4.9. SUBSTRATE PARAMETERS (AQUEOUS ETHANOL)

	Uncharged acids			Charged acids	
HA	m_A	pK_0†	B	m_B	pK_0†
Benzoic	1·57	4·20	Dimethylaniline	5·79	5·03
Isovaleric	1·51	4·78	Trimethylamine	5·43	9·78
Butyric	1·35	4·82	Methylaniline	4·94	4·88
Propionic	1·26	4·87	Dimethylamine	4·86	10·56
Glutaric	1·07	4·34	Methylamine	4·62	10·69
Chloroacetic	1·06	2·86	Ethylamine	4·37	10·72
Acetic	1·04†	4·76	o-Toluidine	4·13	4·47
Succinic	1·02	4·13	m-Toluidine	4·11	4·77
Glycollic	0·94	3·83	Aniline	3·92‡	4·63
Salicylic	0·93	3·00	p-Toluidine	3·81	5·11
Lactic	0·86	3·86	Ammonia	3·79	9·24
Cyanoacetic	0·85	2·47			
Formic	0·77	3·75			
Malonic	0·72	2·75			

† pK value in aqueous solution. ‡ Value obtained through equation 4.27.

Even regardless of sign, the two scales of solvent parameters are quite different. The major change in the magnitude of the Y_- values occurs in the water-rich region from 0 to ca. 60% aqueous ethanol, whereas the Y_0 values change most markedly in the water-poor media from 80% aqueous ethanol to the pure solvent. This means that the solvent effect upon *relative* acid strength is very dependent upon the type of acid–base reaction. The solvent effect upon *absolute* acid strength depends not only upon the Y_- and Y_0 values but also upon $\log f_{H^+}$. This factor also changes little in the water-rich solvents compared to its large value between 80% aqueous ethanol and pure ethanol. The result is that $(pK_s - pK_0)_{HA}$ steadily increases as ethanol content increases, until the sharp increase occurs beyond the 80% solvent. Conversely $(pK_s - pK_0)_B$ decreases with ethanol content passing through a minimum near the 80% composition and thereafter increasing and for some substrates becomes positive in the pure solvent.

Two simple relationships account for these solvent parameters

$$Y_- \simeq (1 - w^2) \tag{4.28}$$

$$Y_0 \simeq -(1 - w)^2 \tag{4.29}$$

where w is the weight fraction of water in the solvent.

Marshall and Grunwald (1954) discovered relationship 4.29 after the examination of $(pK_s - pK)_B$ for aqueous dioxan. The solvent effect on relative acid strength was fitted to equation 4.26 with high precision by using $Y_0 = 0.00$ for water and $Y_0 = -0.635$ for 82% w/w aqueous dioxan the latter calculated from equation 4.29. The procedure appears justifiable in view of the success of this relationship as indicated in Table 4.10. Equation 4.28 is somewhat less satisfactory.

TABLE 4.10. Y_0 VALUES

Solvent % w/w	$-(1-w)^2$	Aq. ethanol	Aq. dioxan	Aq. methanol
0	0.00	0.00	0.00	0.00
20	-0.04	-0.00	-0.05	-0.5
35	-0.12	-0.13$_5$	..	..
45	-0.20	..	-0.23	-0.22$_5$
50	-0.25	-0.26$_5$	..	..
65	-0.42	-0.38	..	..
70	-0.49	..	-0.52	-0.49$_5$
80	-0.64	-0.57	..	..
82	-0.67	..	-0.63$_5$	-0.66
95	-0.90	-0.88	..	..
100	-1.00	-1.00	..	-1.00

Unfortunately this method of scaling Y_0 values for different binary solvents requires that for any pure non-aqueous solvent, S

$$\log(f_{BH^+}/f_B) = m_B \qquad (4.30)$$

which is independent of the nature of S. This is of course not true, and the inevitable result is that the m values become dependent upon the non-aqueous component of the solvent. This is illustrated in Table 4.11.

TABLE 4.11. VARIATIONS IN m_B VALUES

B	Aq. ethanol $m_B - m_{B_0}$	Aq. dioxan $m_B - m_{B_0}$	B	Aq. ethanol $m_B - m_{B_0}$	Aq. dioxan $m_B - m_{B_0}$
Dimethylaniline	1·87	1·34	Aniline	0·00	0·00
Trimethylamine	1·51	−0·40	p-Toluidine	−0·11	−0·19
Methylaniline	1·02	0·76	Ammonia	−0·13	−1·65

Further, the different sets of m values are not simply related to one another by a proportionality factor such as $c = (m_B - m_{B_0})_s/(m_B - m_{B_0})_{s'}$ by which means the Y_0 values could be commonly sealed, i.e. $Y_0 = cY_0'$, with the m values constant. Apparently the same type of situation exists here as in the correlation of solvolysis rates. The structural variations in the acid and its conjugate base are so severe that correlations can only be extended over several binary solvents by allowing variable substrate parameters.

Nevertheless one may examine the m values appropriate to a particular binary solvent and seek to account for their relative magnitudes. This can be done qualitatively for the m_A values in terms of hydrogen-bond solvation.

All the acids HA listed in Table 4.9 are either stronger or have about the same strength as acetic acid, HA_0, in aqueous solution. It follows from equation 4.23 that if $m_A - m_{A_0}$ is negative then, K_A/K_{A_0} in ethanol $> K_A/K_{A_0}$ in water, and the solvent water has exerted a levelling effect on acid strength. This is the expected result for acids stronger than acetic acid if hydrogen-bond formation is more pronounced in water than in ethanol.

It arises

$$HA + CH_3CO_2^- \rightleftharpoons A^- + CH_3CO_2H$$

$$\qquad (4.31)$$

$$OH_2 \qquad HOH \quad HOH \qquad OH_2$$

from the fact that hydrogen bonds are stronger on the left-hand side of equation 4.31, HA is a stronger acid than CH_3CO_2H and $CH_3CO_2^-$ is a stronger base than A^-. This opposes the proton-transfer equilibrium, reduces the observed acidity differences between HA and CH_3CO_2H and accounts for the main trend in the m values given in Table 4.9.

In addition to the above factor, one needs also to consider the solvation of substituents, internal hydrogen-bonding effects and steric interference with solvation.

If HA contains a protic substituent that can be hydrogen bonded to water molecules in the sense of equation 4.32

$$AH + CH_3CO_2^- \rightleftharpoons A^- + CH_3CO_2H$$

$$\begin{array}{ccc} \vdots & & \vdots \\ OH_2 & & OH_2 \end{array} \qquad (4.32)$$

then since AH is more acidic than A^- the HA $\ldots$ OH_2 bond will be stronger than the $A^- \ldots OH_2$ bond, so that again $(K_A/K_{A_0})_{EtOH} > (K_A/K_{A_0})_{H_2O}$ and $m_A - m_{A_0}$ will be negative. This may be a factor in the case of glycollic and lactic acids.

The converse situation applies when HA contains a donor substituent. In this case (equation 4.33)

$$AH + CH_3CO_2^- \rightleftharpoons A^- + CH_3CO_2H$$

$$\begin{array}{ccc} \vdots & & \vdots \\ HOH & & HOH \end{array} \qquad (4.33)$$

the more basic anion will be more strongly hydrogen-bonded, and hence $(K_A/K_{A_0})_{H_2O} > (K_A/K_{A_0})_{EtOH}$ and $m_A - m_{A_0}$ will be positive. This may account for the at first sight anomalous m value for chloroacetic acid and for some increase in m for cyanoacetic acid. There will be a balance between this and the previous factors in most other cases.

Internal hydrogen-bond formation in the anion will tend to "protect" the acid HA from the reduction in ionization constant that occurs when water is

replaced by ethanol as the solvent. This yields $(K_A/K_{A_0})_{EtOH} > (K_A/K_{A_0})_{H_2O}$ and $m_A - m_{A_0}$ negative, as in the case of salicylic and malonic acids.

Superimposed upon these factors there appears to be another associated with the bulk of the hydrocarbon chain in RCO_2H. When R is small, as in formic acid, m is small. The largest values of m are observed for the "large" acids, e.g., benzoic, isovaleric and presumably also glutamic acid. Steric hindrance to anion solvation might be expected to be more serious in water than in ethanol solution, yielding $(K_{HA}/K_{HA_0})_{EtOH} > (K_{HA_0})_{H_2O}$ and hence $m_{HA} < m_{HA_0}$, which is the converse to the observed order. In Chapter 2 it was concluded that, in the absence of chelation effects, substituents close to the carboxyl group do not exert a particularly pronounced steric effect, since relative acid strengths appear to be determined almost completely by polar effects. The observed order $m_{HA} > m_{HA_0}$ is thus unlikely to be due to a greater steric interference with anion solvation in the case of ethanol. There may be another explanation. It appears that the solubility in water of the "large" acids relative to acetic acid is much less than the relative solubilities in ethanol. The reduction in relative anion solubility appears to be substantially smaller, so that, other things being equal, $(K_{HA}/K_{HA_0})_{H_2O} > (K_{HA}/K_{HA})_{EtOH}$ for the "large" acids yielding $m_{HA} > m_{HA_0}$.

The m_B and pK_0 values listed in Table 4.9 suggest that acid and base strength are not important factors controlling the magnitude of m_B. For aqueous ethanol, m_B decreases in the order of tertiary aromatic > tertiary aliphatic > secondary aromatic > secondary aliphatic > normal aliphatic > normal aromatic > NH_3. The main difference in aqueous dioxan is that m_B for aromatic amines is somewhat smaller, but is very much smaller for aliphatic amines, i.e. tertiary aromatic > secondary aromatic > normal aromatic > tertiary aliphatic $\gg NH_3$.

For the charged acids $m_B > m_{B_0}$ means that $(K_B/K_{B_0})_{EtOH} > (K_B/K_{B_0})_{H_2O}$. The acidity of tertiary anilinium cations tends to be increased in ethanol relative to water, whereas the primary species are less acidic in ethanol. The secondary anilinium cations have essentially the same acid strength in the two solvents. All show an increase in acidity in aqueous ethanol relative to pure water or pure ethanol.

The problem of accounting for solvent effects on the acidity of BH^+, or equivalently, the basicity of B, is much simpler in the case of low polarity, aprotic solvents. Pearson and Vogelsong (1958) have determined equilibrium constants at 25°C for reaction 4.34

$$R_3N + ArOH \rightleftharpoons (R_3NH^+.ArO^-) \tag{4.34}$$

involving a series of amines and 2,4-dinitrophenol (ArOH) in several solvents. The products of these acid–base reaction are certainly ion pairs. Table 4.12 lists the log K values obtained.

For the tertiary amines, log K follows dielectric constant reasonably well, as expected, since increasing D will decrease the electrostatic free energy of the ionic species. For each solvent, the same order of base strength $(C_2H_5)_3N > (C_4H_9)_3N > (CH_3)_3N$ is observed, and this may be the order

TABLE 4.12. AMINE-2,4-DINITROPHENOL ION PAIR FORMATION (25°C)

Amine	C_7H_{16} 1·9†	Dioxan 2·2†	C_6H_6 2·3†	$CH_3CO_2CH_3$ 6·3†	C_6H_5Cl 5·6†	$CHCl_3$ 4·8†	$C_2H_4Cl_2$ 10·2†
$(C_4H_9)_3N$	2·0	2·63	2·79	3·40	3·58	3·96	4·47
$(C_2H_5)_3N$	..	3·16	3·49	4·08	3·99	4·20	..
$(CH_3)_3N$	..	2·58	2·90	..	3·14	3·43	..
$(C_4H_9)_2NH$	2·56	3·60	2·98	4·56	3·40	3·41	4·08
$(C_2H_5)_2NH$	..	3·79	3·09	4·68	3·40	3·24	..
$(CH_3)_2NH$	..	3·73	2·94	..	3·03	3·09	..
$C_6H_{13}NH_2$	1·70	3·63	1·72	4·70	1·90	1·72	..
$C_4H_9NH_2$	..	3·58	..	4·58	..	1·65	..
$C_2H_5NH_2$	..	3·69	..	4·82	..	..	..
CH_3NH_2	..	3·68	..	..	..	1·54	..

† Dielectric constant (D).

of "intrinsic" base strengths. The addition of up to 5 mole % of ethanol has no detectable effect on the K value for $(CH_3)_3N$ in $CHCl_3$.

Much the same trend, but with more variations is observed for the secondary amines, with the exception of solutions in dioxan or ethyl acetate. In these weakly basic solvents the K values are distinctly larger than expected from considerations of dielectric constant. Unlike the tertiary ammonium cation, where the single proton will be closely associated with the anion, there is a second proton in the secondary cation.

$$R_3\overset{+}{N}-H \cdots \overset{-}{O}Ar \qquad R_2\overset{+}{N}-H \cdots \overset{-}{O}Ar$$
$$\underset{H \cdots \text{Solvent}}{|}$$

The additional hydrogen-bonded interaction between the cation and the basic solvent molecule will increase the relative stability of the ion-pair product. Addition of ca. 4 mole % ethanol to diethylamine in chloroform increases K by a factor of about 10.

The basic solvents are superior to non-basic solvents of similar dielectric constant by about two powers of ten for ion-pair formation from the primary amines. There are now three protons on nitrogen to be solvated. In chloroform solution addition of ca. 5 mole % ethanol to methylamine leads to a 300-fold increase in K. Similarly ca. 3 mole % ethanol increases K for benzylamine by a factor of 30. Additions of ca. 3 mole % of either diethyl ether or dioxan to benzylamine in chloroform increases K by a factor of 3.

For the primary amines, a second factor is now apparent in that chloroform appears to lead to lower K values than expected. This can arise from solvent stabilization of the amine by weak hydrogen bonding, $N \cdots H—CCl_3$. This explanation is supported by the partition coefficients between water and various solvents for these amines (see Table 4.13).

In all the solvents listed in Table 4.13 except $CHCl_3$, log p for each amine has about the same value. This value appears to depend upon the number of $N—H \cdots O$ bonds that can be formed in water. Compared with the other solvents, chloroform is found to be a superior solvent for the amines. In fact it is a better solvent than water in the tertiary amine case.

TABLE 4.13. PARTITION COEFFICIENTS OF AMINES

log p = log[Amine in water]/[Amine in solvent]

	$CHCl_3$	CCl_4	C_6H_5Cl	C_6H_6	$(C_2H_5)_2O$
$(CH_3)_3N$	−0·35	0·11	0·30	0·32	0·34
$(CH_3)_2NH$	0·44	1·13	1·13	1·05	1·26
CH_3NH_2	1·06	1·49	1·76	1·44	1·64
NH_3	1·42	2·39	2·25	2·32	2·26

If some restrictions are placed upon the extent of variations in substrate structure, then these variations may be correlated by the Hammett or Taft equations. For example, if in equation 4.23 HA is a substituted acetic acid and HA_0 acetic acid itself then

$$(pK_A - pK_{A_0})_s - (pK_A - pK_{A_0}) = (m_A - m_{A_0})Y_-$$
$$= \sigma_A^*(\rho_{a,0}^* - \rho_{a,s}^*) \qquad (4.34)$$

where σ_A^* is the appropriate Taft substituent parameter ($\sigma_{A_0}^* = 0$), $\rho_{a,s}^*$ is the reaction parameter appropriate to the solvent S.

It follows from equation 4.34 that

$$\frac{(\rho_{a,0}^* - \rho_{a,s}^*)}{Y_-} = \frac{m_A - m_{A_0}}{\sigma_A^*} = q_-^* \qquad (4.35)$$

where q_-^* must be independent of substituent and solvent being determined only by the reaction taking place, i.e. ionization of aliphatic carboxylic acids, and the temperature. A similar equation, 4.36, can be developed for solvent effects on the acidity of substituted methylammonium cations

$$\frac{(\rho_{b,0}^* - \rho_{b,s}^*)}{Y_0} = \frac{(m_B - m_{B_0})}{\sigma_B^*} = q_0^* \qquad (4.36)$$

For substituted benzoic acids and anilinium cations the equations equivalent to 4.35 and 4.36 are

$$\frac{(\rho_{a,0} - \rho_{a,s})}{Y_-} = \frac{(m_{A,R} - m_{A,H})}{\sigma_R} = q_- \qquad (4.37)$$

$$\frac{(\rho_{b,0} - \rho_{b,s})}{Y_0} = \frac{(m_{B,R} - m_{B,H})}{\sigma_R} = q_0 \qquad (4.38)$$

Where $m_{A,H}$ and $m_{B,H}$ are the m_A and m_B values for benzoic acid and the anilinium cation, respectively, and $m_{A,R}$ and $m_{B,R}$ refer to the substituted compounds.

Equation 4.37 shows that the Y_- values can be used to account for the variation of reaction parameters with solvent. This may be expressed as

$$\rho_{a,s} = \rho_{a,0} - q_- . Y_- \qquad (4.39)$$

By using ρ values based only upon "well behaved" m-substituents for water (1·00), 50% aq. ethanol (1·57) and ethanol (1·61), a value $q_- = 0·65 \pm 0·05$ can be obtained. This has been employed to estimate the Y_- values listed in Table 4.14. Again the ρ values are based upon m-substituents only.

Table 4.14. Y_- VALUES FROM VARIATIONS IN ρ VALUES

Solvent	ρ	Y_-	Solvents	ρ	Y_-
Methanol	1·58	0·89	26·5% w/w aq. dioxan	1·18	0·28
n-Propanol	1·63	0·97	43·5% w/w aq. dioxan	1·26	0·40
n-Butanol	1·50	0·77	73·5% w/w aq. dioxan	1·42	0·65
Ethylene glycol	1·39	0·60	Benzene	2·15	1·8

For benzene solution, the equilibrium actually observed by Davis and Hetzer (1958) was the relative formation constants of ion pairs with diphenylguanidine. These data may be regarded as yielding measures of "intrinsic" substituent effects free from specific solvation phenomena. From a comparison of relative acid strength in benzene and the other hydroxylic solvents, Davis and Hetzer found that the m-OH group displays a marked dependence on solvent composition, being substantially acid strengthening only in benzene, water and aqueous solvents. The p-OH and, to a lesser extent, the p-OCH$_3$ and p-F groups are always more acid weakening in hydroxylic solvents than in benzene, whereas the p-NO$_2$ group appears to be somewhat more acid strengthening in aqueous media. These findings have of course been taken into account in the σ^0 values (see Table 2.4).

All o-substituents are substantially less acid strengthening, relative to m- and p-substituents, in benzene solution than they are in the hydroxylic solvents. This suggests that a large part of the acid-strengthening effect of o-substituents in hydroxylic solvents arises from steric hindrance to solvation of the carboxylic acid group and not to steric inhibition of resonance. This will arise if the precise orientation of solvent molecules about its carboxyl group is more critical than that about the carboxylate anion.

Corresponding to equation 4.39, the effect of solvent change on the reaction parameters for the anilinium cations will be given by

$$\rho_{b,s} = \rho_{b,0} - q_0 Y_0 \tag{4.40}$$

In order to account for the observed ρ values for water and for ethanol solution (equations 2.94 and 4.50, respectively), q_0 must be given the value 1·56.

4.4 ACIDITY FUNCTIONS

The acids and bases discussed in the previous Section are all of such strength that ionization constants could be determined in extremely dilute aqueous solutions. Under these conditions, the concentrations are excellent approximations for the activities of the species in equilibrium, and a satisfactory extrapolation to infinite dilution can be made to obtain the thermodynamic ionization constant K. In order to obtain measurable quantities of conjugate acid and base in equilibrium for very weak acids or bases, solutions of high basicity or acidity are required. Such solutions are at least as far from the ideal of infinitely dilute aqueous solution as are the mixed-solvent media described above.

The measurable quantity from which one hopes to derive the pK of the conjugate acid of a very weak base B is the concentration ratio, $(BH^+)/(B) = \varphi$, determined, for example, spectrophotometrically with allowances for medium effects on the spectrum. Equation 4.41 indicates the quantities required in the non-ideal case.

$$pK = -\log a_{H^+} + \log(f_{BH^+}/f_B) + \log \varphi \tag{4.41}$$

The f's are activity coefficients relative to infinitely dilute aqueous solutions ($f = 1$) and a_{H^+} is the activity of protons in the medium. If the quantity

$$H_0 = -\log a_{H^+} + \log(f_{BH^+}/f_B) \tag{4.42}$$

were known for the medium, then the pK could be determined from equation 4.41.

Hammett and Deyrup (1932) approached this problem by suggesting that for a series of structurally related indicator bases, the activity-coefficient ratios, $\log(f_{BH^+}/f_B)$, might be constant. If so, then the function defined by equation 4.42 would be a constant for a given acidic medium. This is termed the H_0 *acidity function*, alternatively symbolized by h_0 where $-\log h_0 = H_0$.

The H_0 function was determined by first selecting a base whose pK value could be determined in dilute aqueous solution. From

$$H_0 = pK - \log \varphi \qquad (4.43)$$

the H_0 values for progressively more acidic solutions can be determined until the ratio $(B)/(BH^+)$ becomes too small to be accurately measured. The usual range of measurable $\log \varphi$ is -1.5 to $+1.5$. The pK of a somewhat weaker base can then be obtained using these H_0 values and then this base employed to extend H_0 into more acidic solution. By means of a series of bases with overlapping useful regions the whole H_0 scale was set up.

The H_0 values are listed in Table 4.15 for fairly strongly acidic solutions (60% w/w $H_2SO_4 \equiv 9.2M$). For nitric, phosphoric, hydrochloric and hydrobromic acids, a $0.1M$ solution is almost ideal, but at this concentration sulphuric acid is stronger ($H_0 < -\log H^+$) than expected. For all of these acids, the acidity as measured by H_0 increases far more rapidly the $\log H^+$, even for relatively weak phosphoric acid at $> 5M$. Hydrogen fluoride at the $10M$ level appears substantially more acidic than at the M level.

TABLE 4.15. H_0 VALUES (25°C)

(H^+)†	$-\log(H^+)$†	$HClO_4$	H_2SO_4	HNO_3	H_3PO_4‡	HBr	HCl	HF
0·1M	1·00	..	0·83	0·98	1·45	0·98	0·98	..
0·5M	0·30	0·18	0·13	0·21	0·97	0·20	0·20	..
M	0·00	−0·22	−0·26	−0·18	0·63	−0·20	−0·20	1·20
2M	−0·30	−0·78	−0·84	−0·67	0·24	−0·71	−0·69	0·91
5M	−0·70	−2·23	−2·28	−1·57	−0·69	−1·93	−1·76	0·28
7M	−0·85	−3·61	−3·32	−1·99	−1·45	−2·85	−2·56	0·02
10M	−1·00	−5·79	−4·89	..	−2·59	−4·44	−3·68	−0·36

† Assuming complete dissociation. ‡ At 19°C.

The method of derivation of the H_0 values depends upon the postulate that for a series of "Hammett" bases B_1, B_2, etc.

$$\log[f_{B_1H^+}/f_{B_1}] = \log[f_{B_2H^+}/f_{B_2}] = c \qquad (4.44)$$

where c is a constant dependent upon the medium, but independent of the base. This is essentially the activity postulate, equation 4.25, in a more restrictive form, i.e.

$$c = m_{B_1} Y_0 = m_{B_2} Y_0 = m_{B_3} Y_0 \text{ etc.} \qquad (4.45)$$

The test of the postulate is that for some base, B′, a plot of its indicator ratio, $\log \varphi'$, against the H_0 function, based upon a similar base B, shall be linear with slope of $-1\cdot00$. However the less restrictive activity postulate 4.25 will lead to

$$\log \varphi' = pK' - H_0 + (m_B - m'_B)Y_0 \tag{4.46}$$

which gives the same result, provided that the term containing Y_0 remains constant within the range of variation of H_0. The m and Y_0 values listed in Tables 4.9 and 4.10 indicate that this term is probably quite small in water-rich media for bases of the same type, i.e. all primary aromatic amines.

For solutions up to ca. 65% w/w sulphuric acid the H_0 function is based almost completely upon the results for nitroanilines. In higher acid concentration, a tertiary amine and several ketones were employed. Certainly in the case of tertiary amines equation 4.44 is not satisfied.

Several bases have been examined to determine whether or not they can be classified as "Hammett" bases, i.e. whether the activity coefficient ratios in equation 4.44 vary with acidity in the same way as those for the nitroanilines. Some aromatic ethers, carboxylic acids, aldehydes, ketones, simple amides and azo-compounds seem to be in this class. Secondary and tertiary amines, some o-substituted benzoic acids and aliphatic amides behave differently. Different acidity functions can be based upon these indicators.

Triarylcarbinols display completely different behaviour on protonation (equation 4.47)

$$ROH + H^+ = R^+ + H_2O \tag{4.47}$$

and form the basis for the H_R function defined by

$$H_R \equiv pK_R - \log[(R^+)/(ROH)]$$
$$= -\log a_{H^+} + \log a_{H_2O} + \log(f_{R^+}/f_{ROH}) \tag{4.48}$$

The difference between the H_R and H_0 functions arises not only from the presence of the $\log a_{H_2O}$ term in the former, but also from differences between f_{R^+}/f_{ROH} and f_{BH^+}/f_B.

Other acidity functions have been defined which are appropriate to other types of acid–base pair. Thus

For uncharged acids

$$HA = H^+ + A^-$$
$$H_- \equiv pK_{HA} - \log[(HA)/(A^-)]$$
$$= -\log a_{H^+} + \log(f_{HA}/f_{A^-}) \tag{4.49}$$

For cationic bases

$$BH^{++} = B^{+} + H^{+}$$

$$H_{+} \equiv pK_{BH^{++}} - \log[(BH^{++})/(B^{+})]$$

$$= -\log a_{H^{+}} + \log f_{BH} \qquad (4.50)$$

Very little work has been reported concerned with the H_{+} function but the H_{-} function has obvious importance. Values have been listed for aqueous alkali hydroxides, aqueous hydrazine, aqueous ethylenediamine and methanolic sodium methoxide. Unfortunately, various indicator acids have been employed that do not obey equation 4.44. Employing the strongly acidic cyanocarbon acid indicators, e.g. pentacyanopropene and bis(tri-cyanovinyl)amine, Boyd (1963) determined an H_{-} function for aqueous sulphuric acid. These results are compared in Table 4.16 with the corresponding H_0 and H_R functions.

TABLE 4.16. COMPARISON OF ACIDITY FUNCTIONS

Acid % w/w	Aq. sulphuric acid			Aq. perchloric acid	
	H_0	H_R	H_-	H_0	H_R
10	-0.3	-0.7	-0.1	-0.2	-1.0
25	-1.4	-2.6	-1.5	-1.0	-3.1
50	-3.4	-6.2	-3.9	-2.2	-7.9
60	-4.5	-8.1	-4.9	-2.8	-11.1

The H_0 function for highly concentrated aqueous sulphuric acid solutions, 83–99.8% w/w sulphuric acid, can be accounted for by means of two surprisingly simple expressions, 4.51 due to Brand (1950) and 4.52 due to Deno and Taft (1954).

$$H_0 = -8.36 + \log[(HSO_4^-)/(H_2SO_4)] \qquad (4.51)$$

$$= -6.66 + \log[(H_2O)/(H_3O^+)] \qquad (4.52)$$

where the molar concentrations, e.g. HSO_4^-, are calculated from the stoicheiometric amounts of sulphuric acid and water and an estimated equilibrium constant of 50, ($\log 50 = 1.70$), for the reaction

$$H_2SO_4 + H_2O \rightleftharpoons H_3O^+ + HSO_4^-$$

Equations 4.51 and 4.52 may be compared with equation 4.43, and are seen to require that HSO_4^- and H_2O function as Hammett bases in these solutions. In other words

$$\frac{f_{HSO_4^-} f_{BH^+}}{f_{H_2SO_4} f_B} = \frac{f_{H_2O} f_{BH^+}}{f_{H_3O^+} f_B} = \text{constant} \tag{4.53}$$

Equation 4.53 is certainly not true in more dilute sulphuric acid solutions, and, since the constancy of these ratios requires the cancellation of activity-coefficient ratios involving species of different charge type, it appears necessary that the individual activity coefficients have become constant. Deno and Taft suggest that the reason for the ideal behaviour may be a result of the high dielectric constant of sulphuric acid (110), the similarity of the oxygens of H_2SO_4 and HSO_4^- and the resemblances between this and fused-salt systems which also appear to be ideal.

The activity, and hence activity coefficient of water, in these media can be calculated from vapour-pressure data, and it is found that in $> 83\%$ sulphuric acid log f_{H_2O} remains constant at -1.67. In more dilute solutions $-\log f_{H_2O}$ decreases and if allowance for this factor is made (equation 4.54), the calculation of H_0 can be extended down to ca. 75% w/w sulphuric acid solution.

$$H_0 = -6.66 + \log[(H_2O)/(H_3O^+)] + \log f_{H_2O} \tag{4.54}$$

From the point of view of water molecules solvating the various species, the solutions considered above are essentially dry. It is not until the sulphuric acid content falls below 60% that there is a significant quantity of water available for solvation purposes.

For the region 45–60% w/w sulphuric acid, Taft (1960) has observed that the difference between the H_R and H_0 functions varies linearly with the activity of water. For an acidity function, H_0', based on primary amines, $H_R - H_0'$, is proportional to $-4 \log a_{H_2O}$, whereas for a function based on secondary amines, H_0'', the plot of $H_R - H_0''$ versus $-\log a_{H_2O}$ has a slope of 3. This behaviour is most readily accounted for in terms of the number of water molecules hydrogen bonded to the acidic species. If the relevant equilibria are written as

$$R^+ + (n+1)H_2O = ROH + H^+ . nH_2O$$

$$B_1H^+ . n_1H_2O + (n-n_1)H_2O = B_1 + H^+ . nH_2O$$

$$B_2H^+ . n_2H_2O + (n-n_2)H_2O = B_2 + H^+ . nH_2O$$

then

$$H_R \equiv -\log a_{H^+} + (n+1) \log a_{H_2O} + \log(f_{R^+}/f_{ROH})$$

$$H_0' \equiv -\log a_{H^+} + (n-n_1) \log a_{H_2O} + \log(f_{BH^+}/f_B)$$

$$H_0'' \equiv -\log a_{H^+} + (n-n_2) \log a_{H_2O} + \log(f_{BH^+}/f_B)$$

and

$$H_R - H'_0 = (n_1 + 1)\log a_{H_2O} + \log(f_{R^+} f_B / f_{BH^+} f_{ROH}) \qquad (4.55)$$

$$H_R - H''_0 = (n_2 + 1)\log a_{H_2O} + \log(f_{R^+} f_B / f_{BH^+} f_{ROH}) \qquad (4.56)$$

The solvated cations may be formulated as

$$
\begin{array}{ccc}
\text{H} \cdots \text{OH}_2 & \text{H} \cdots \text{OH}_2 & \text{H} \cdots \text{OH}_2 \\
| & | & | \\
\text{Ar}-\overset{+}{\text{N}}-\text{H} \cdots \text{OH}_2 & \text{Ar}-\overset{+}{\text{N}}-\text{H} \cdots \text{OH}_2 & \text{O}^+-\text{H} \cdots \text{OH}_2 \\
| & | & | \\
\text{H} \cdots \text{OH}_2 & \text{R} & \text{H} \cdots \text{OH}_2 \\
n_1 = 3 & n_2 = 2 & n = 4
\end{array}
$$

Extrapolation of the linear relationships indicates that the values of ~ -2 for $\log(f_{R^+} f_{B_2}/f_{B_2H^+} f_{ROH})$ and ~ -1.5 for $\log(f_{R^+} f_{B_1}/f_{B_1H^+} f_{ROH})$ must have remained constant throughout the composition range examined.

In the range 62–74% w/w sulphuric acid, $H_R - H'_0$ versus $-\log a_{H_2O}$ is not linear. The apparent slope $(n_1 + 1)$ falls from 3 to 1.8. In the same range, $H_R - H'''_0$, based on a tertiary amine the apparent slope $(n_3 + 1)$ falls from 1.7 to ca. 1. These results may be interpreted as indicating that the concentration of water is now so low that solvation of the anilinium cations is incomplete.

It is not necessary to assume, as implied above, that the uncharged bases ROH, B_1 and B_2 are not solvated, since provided the number of water molecules associated with each of these species is the same, then this number will not appear in equations 4.55 and 4.56. Several pieces of evidence suggests that the triarylmethyl cations, R^+, are not specifically solvated by the hydrogen-bond formation envisaged for the protic cations. The formation of a similar aprotic cation NO^+ from HNO_2 follows the H_R function.

A proton hydration number of 4 is also suggested by the results of Bell (1959). It will be noted that in Table 4.15 there appears to be a marked dependence of H_0 upon the anion of the acid concerned. This dependence is no longer evident when H_0 values at equal molalities instead of molar concentrations are compared as in Table 4.17. The calculated values of H_0 are obtained from equation 4.57, which is based on the assumption that the solvated proton is $H_9O_4^+$

$$-H_0 = \log m - 4\log(1 - 0.072\,m) + 3\log(1 + 0.032\,m) \qquad (4.57)$$

Arnett and Mach (1964) have re-examined the relationship between the H_R, H'_0, H'''_0 and also the H'_R function, which is based upon the protonation of diarylolefines (O)

$$R^+ + nH_2O = O + H^+ \cdot nH_2O$$

TABLE 4.17. H_0 VALUES AND MOLALITY

(Molality, m = moles acid/1000 grams water)

Molality	HClO$_4$	H$_2$SO$_4$	HBr	HCl	Average $-H_0$	Calculated $-H_0$
1·0	0·19	0·24	0·18	0·19	0·20	0·17
2·0	0·64	0·76	0·67	0·64	0·68	0·64
6·0	2·00	1·99	2·01	1·89	1·99	1·99
10·0	3·41	3·18	3·31	2·97	3·57	3·57

They conclude that the various $H_R - H_0$, $H'_R - H_0$ and $H'''_0 - H'_0$ versus log a_{H_2O} plots are not linear, and that the apparent hydration numbers, n_1 and n_3, obtained by drawing tangents to the curves correspond to large variations in hydration with acid concentration between 10 and 90% w/w sulphuric acid. In the 40–60% range, n_1 varies from 4 to 2 and n_3 from 3 to 1. In 10% w/w sulphuric acid, the apparent values are $n_1 \sim 13$ and $n_3 \sim 6$. When these results and the activity coefficient data on the bases and their conjugate acids in sulphuric acid due to Boyd (1963) are compared, it is apparent that the relevant activity-coefficient ratios are sensitive to substrate structure and acid concentration. The most pessimistic view is that Taft's assessment of hydration numbers may result from an artifact. Nevertheless, the results do appear consistent with the general qualitative theory of solvent effects on the acidity of protonated amines described in Section 4.3.

The doubts that have cast upon the concept of acidity functions undermine the two main uses to which they have been put (cf. Paul and Long, 1957). The most obvious is the determination of otherwise inaccessible pK values. The second use is in reaction-mechanism studies.

There are many reactions that are subject to acid catalysis. The function of the catalyst is to convert a reactant (X) to a more reactive species, and since, by definition, an acid is a proton donor, this species must be the conjugate acid (XH$^+$) of the reactant or species obtained from it. Many of these reactions have as a rate-controlling step either the unimolecular decomposition of XH$^+$ (A1 mechanism) or its bimolecular reaction with solvent water (A2 mechanism). These two processes are kinetically indistinguishable, but it is sometimes possible to draw mechanistic conclusions on the basis of correlation of rate constants with the H_0 function.

Representing the A1 mechanism by

$$X + H^+ \rightleftharpoons XH^+; \quad XH^+ \xrightarrow{1} \text{Products}$$

$$\log k_{obs} = \log(k_1/K_{BH^+}) + \log a_{H^+} + \log(f_X/f_1^{\ddagger}) \qquad (4.58)$$

where k_{obs} is the observed first-order rate constant, and $f_i^{\ddagger}$ is the activity coefficient of the transition state of the unimolecular step.

Equation 4.58 is equivalent to

$$\log k_{obs} = -H_0 + \log(f_X f_{BH^+}/f_i^{\ddagger} f_B) + \text{constant} \qquad (4.59)$$

Zucker and Hammett (1939) suggested that for this mechanism, $\log k_{obs}$ will vary linearly with H_0 having slope $-1{\cdot}00$ i.e. $k_{obs} = kh_0$, if the activity coefficient ratios are constant. This might well be the case if X is a "Hammett" base, since the transition state should closely resemble XH^+. For most reactions, however, X is very different from the bases used to set up H_0, and often plots of $\log k_{obs}$ versus H_0 are non-linear or are approximately linear with slopes different from $-1{\cdot}00$.

Representing the A2 mechanism by

$$X + H^+ \rightleftharpoons XH^+; \quad XH^+ + H_2O \xrightarrow{2} \text{Products}$$

$$\log k_{obs} = \log(k_2/K_{BH^+}) + \log a_{H^+} + \log a_{H_2O} + \log(f_X f_{H_2O}/f_2^{\ddagger})$$
$$= \log H^+ + \log(f_X f_{H^+}/f_2^{\ddagger}) + \log a_{H_2O} + \log(k_2/K_{BH^+}) \qquad (4.60)$$

Equation 4.60 suggests that no parallel between $\log k_{obs}$ and H_0 will be observed but k_{obs} may be proportioned to the molar concentration of acid if $\log(f_X f_{H^{\ddagger}}/f_2^{\ddagger}) + \log a_{H_2O}$ happens to be constant. This is rarely observed, although there are a few examples of reactions classified as A2 for which $k_{obs} \simeq kH^+ \neq kh_0$.

Frequently observed, unfortunately, is the ambiguous result that k_{obs} varies in some intermediate manner between h_0 and H^+. Further, if the protonation of X (k_i) is the rate-controlling step then

$$\log k_{obs} = \log k_i + \log a_{H^+} + \log(f_X/f_i^{\ddagger})$$
$$= -H_0 + \log(f_X f_{BH^+}/f_i^{\ddagger} \cdot f_B) + \text{constant} \qquad (4.61)$$

and again $\log k_{obs}$ may be approximately linear with H_0.

More definite information may be obtained when the mechanistic distinction arises from a difference between the H_0 and H_R functions. The outstanding example in this case is the distinction between $H_2NO_3^+$ and NO_2^+ as the effective reagent in nitration. Since the rate of nitration parallels the H_R rather than the H_0 function, the nitrating agent clearly arises by

$$HNO_3 + H^+ \rightarrow NO_2^+ + H_2O$$

and not by

$$HNO_3 + H^+ \rightarrow H_2NO_3^+$$

REFERENCES

Arnett, E. M., Duggleby, P. M., and Burke, J. J. (1963). *J. Am. chem. Soc.*, **85**, 1350.
Arnett, E. M., and Mach, G. W. (1964). *J. Am. Chem. Soc.*, **86**, 2671.
Bell, R. P. (1959). "The Proton in Chemistry", Cornell, University Press, Ithaca, New York.
Boyd, R. H. (1963). *J. Am. chem. Soc.*, **85**, 1555.
Brand, J. C. D. (1950). *J. chem. Soc.*, 1002.
Davis, M. M., and Hetzer, H. B. (1958). *J. Res. natn. Bur. Stand.*, **60**, 569.
Deno, N. C., and Taft, R. W. (1954). *J. Am. chem. Soc.*, **76**, 244.
Fainberg, A. H., and Winstein, S. (1956). *J. Am. chem. Soc.*, **78**, 2770.
Gould, E. S. (1960). "Mechanism and Structure in Organic Chemistry", Holt, Reinhart and Winston, New York.
Grunwald, E., and Berkowitz, B. J. (1951). *J. Am. chem. Soc.*, **73**, 4939.
Grunwald, E., and Winstein, S. (1948). *J. Am. chem. Soc.*, **70**, 846.
Gutbezahl, B., and Grunwald, E. (1953). *J. Am. chem. Soc.*, **75**, 559, 565.
Hammett, L. P., and Deyrup, A. J. (1932). *J. Am. chem. Soc.*, **54**, 2721.
Hine, J. (1962). "Physical Organic Chemistry". McGraw-Hill, New York.
Marshall, H. P., and Grunwald, E. (1954). *J. Am. chem. Soc.*, **76**, 2000.
Paul, M. A., and Long, F. A. (1957). *Chem. Rev.*, **57**, 1, 935.
Pearson, R. G., and Vogelsong, D. C. (1958). *J. Am. chem. Soc.*, **80**, 1038.
Ross, S. D., and Labes, M. M. (1957). *J. Am. chem. Soc.*, **79**, 4155.
Swain, C. G., Mosely, R. B., and Bown, D. E. (1955). *J. Am. chem. Soc.*, **77**, 3731.
Taft, R. W. (1960). *J. Am. chem. Soc.*, **82**, 2965.
Winstein, S., and Fainberg, A. H. (1957). *J. Am. chem. Soc.*, **79**, 5937.
Winstein, S., Fainberg, A. H., and Grunwald, E. (1957). *J. Am. chem. Soc.*, **79**, 4146.
Zucker, L., and Hammett, L. P. (1939). *J. Am. chem. Soc.*, **61**, 2785, 2791.

5

Correlation of Reagent Changes

5.1 THE BRØNSTED CATALYSIS EQUATION

The earliest reported LFER is the correlation of acid and base strength with effectiveness as catalysts in reactions subject to general acid–base catalysis. This was developed by Brønsted and Pederson (1924) largely as a result of studies on the decomposition of nitramide. The relationship may be expressed by equations 5.1 and 5.2,

$$\log k_A = \log G_A + \alpha \log K_A \qquad (5.1)$$

$$\log k_B = \log G_B - \beta \log K_A \qquad (5.2)$$

where k_A and k_B are the catalytic constants in some process for acid (A) and base (B) catalysis, respectively; K_A is the dissociation constant of the acid, A, or the conjugate acid of the base, B; and G_A, G_B, α and β are "constants" that depend upon the reaction process and the conditions of temperature and solvent.

For specific acid or base catalysis in the same solvent, namely water, for which K_A has been determined, α and $\beta = 1$. The only observable catalysts in these cases are the conjugate acid or base of the solvent. As α and β tend to zero, catalysis by the solvent itself swamps all other and acid–base catalysis can no longer be demonstrable. In reactions subject to general catalysis, α and β are positive and less than unity.

Statistical corrections are required in correlating the effect of catalysts of differing types. The modified equations 5.3 and 5.4

$$\log(k_A/p) = \log G_A + \alpha \log(qK_A/p) \qquad (5.3)$$

$$\log(k_B/q) = \log G_B - \beta \log(qK_A/p) \qquad (5.4)$$

are often employed, in which p is the number of equally bound dissociable protons in the acid catalyst (or conjugate acid of the base catalyst) and q is the number of equivalent sites for protonation in the base catalyst (or conjugate base of the acid catalyst). The most fundamentally satisfying solution to this problem, discussed by Benson (1958), is use of symmetry numbers for reactants, products and transition states. However, conventionally (cf. Bell, 1941) it is usual to apply the statistical corrections only for protons and basic sites on different, but equivalent, atoms. Thus one takes $p = 1$ for NH_4^+ not 4 and $q = 1$ for RCO_2^-, whereas $p = 2$ for $HO_2C(CH_2)_nCO_2H$ and $q = 2$ for NH_2NH_2. The range of variation of k_A, k_B and K_A and deviations from various sources are usually such that the statistical correction is a relatively trivial factor.

The decomposition of nitramide is both the most studied and the best correlated general base-catalysed reaction. Equation 5.4 is adequate for covering an extremely wide range of bases, but only shows high precision when the catalysts are examined separately according to charge type. Table 5.1, taken from Bell (1958), gives some examples for aqueous solution at 15°C. Within each charge type, the bases are all structurally similar. Thus the B^- class contains fourteen carboxylates, primary phosphate and hydroxide. Bases of type B^{++} are rather poorly correlated.

TABLE 5.1. BRØNSTED CORRELATION OF NITRAMIDE DECOMPOSITION

Base type	Example	Number of catalyst	G_B	β
B^{--}	Succinate	5	2.1×10^{-5}	0·87
B^{++}	$[Co(NH_3)_5OH]^{++}$	7	7.8×10^{-3}	0·82
B^-	Acetate	16	7.2×10^{-5}	0·80
B^0	Aniline	8	1.7×10^{-4}	0·75

The most intensively studied acid-catalysed reaction has been the dehydration of acetaldehyde hydrate. Bell (1959) reports a good Brønsted correlation covering 45 carboxylic acids and phenols having $\alpha = 0.54$.

Many reaction series have been described that follow the Brønsted equation with reasonable precision. Some examples discussed by Bell (1941, 1959) are the mutarotation of glucose and the halogenation of acetone. Sufficient data are seldom available on acids and bases of differing charge and structural types for separate correlations to be attempted.

By selecting some acid (or base) as a standard and considering relative catalytic constants and acidities, equation 5.1, and similarly equation 5.2, can be expressed as a two-parameter LFER,

$$\log(k_A/k_A^0) = \alpha \log(K_A/K_A^0) \tag{5.5}$$

in which $\log(K_A/K_A^0)$ could be represented by some symbol, e.g. R_A, and thought of as a reagent parameter. The reaction parameter, α, then measures the susceptibility of the catalysed process to changes in relative acidity. When the catalysing acids are m- and p-substituted benzoic acids, the reagent parameter, R_A, is simply the Hammett σ value if benzoic acid is the chosen standard. In this case α is the ρ value that would be obtained if the reaction series was viewed in terms of the catalysing acid as the substrate. In the general case of catalysis by aromatic acids or bases in any solvent

$$\alpha \quad \text{or} \quad \beta = \rho_R/\rho_a$$

where ρ_a applies to the ionization of the acids and ρ_R applies to the catalysed reaction viewed as a Hammett series. Similarly, for a series of substituted aliphatic acids

$$\alpha \quad \text{or} \quad \beta = \rho_R^*/\rho_a^*$$

The extent of permissible variation in acid and base structure is much greater for Brønsted correlations than for Hammett and Taft correlations. This arises from the greater restriction in the type of reaction process, i.e. proton transfers. Limitations in structural variations are still important is indicated above in the case of nitramide decomposition. Thus the "pseudo acids" do not fit the Brønsted equation for the dehydration of acetaldehyde hydrate, presumably since the pronounced delocalization present in these anions is not developed in the transition state of the proton-transfer reaction. Imidazole and hydroxylamine are exceptionally good base catalysts for the hydrolysis of phenyl acetate. This, however, is probably not a case of base catalysis but of nucleophilic catalysis.

Since Hammett and Taft correlations of acid strength are obtained for a wide variety of solvents, it is to be expected that the Brønsted equation will still apply to reactions taking place in media other than that for which K_A is determined. Table 5.2 illustrates the parameters obtained for correlations of the nitramide decomposition catalysed by amine bases in various solvents. The K_A values employed are for the conjugate acids in aqueous solution.

No parallel is observed between either the β or the G_B values and dielectric constant, acidity or basicity of the solvents. There is also no parallel between the β values and the G_B values in Tables 5.1 or 5.2. In other words, there is no indication of a "selectivity–activity" relationship, in which a correlation

TABLE 5.2. NITRAMIDE DECOMPOSITION IN VARIOUS SOLVENTS

(Caldin and Peacock, 1955; Callender et al., 1960)

Solvent	Dielectric constant	β	$10^6 G_B$
Water	78·5	0·75	36
Nitrobenzene	34·8	0·67	2·9
m-Cresol	11·8	0·84	17
Isopentanol	5·7	0·92	0·8
Anisole	4·3	0·64	1·1
Benzene	2·3	0·7	1·3

between a substrate selectivity between catalysts, measured by α or β, and its activity, measured by G, the rate constant for a catalyst of unit acidity ($K_A = 1$).

There are several possible reaction mechanisms for which the observation of general acid–base catalysis may be expected. These are discussed by Bell (1959). In the simplest of these, the rate-controlling step is the protonation (or deprotonation) of the substrate by the acid (or base) catalyst. Protonation and deprotonation rates also enter into the expression for k_A or k_B in more complex cases. This suggests that the rates of ionization of acids, i.e. proton transfer to the reference base, water, should be related to the ionization constant. Several slowly ionizing acids for which this rate has been measured show such a parallel. The reaction mechanism presumably involves the formation of hydrogen-bonded complexes, particularly in poorly ionizing media. Gordon (1961) reports a linear relationship between hydrogen-bond formation and acid strength whose result will be that the Brønsted equation will be applicable to these processes.

An important limitation of the Brønsted equation has been recognized by Eigen (1964). As the difference between the acidity of the catalyst and conjugate acid of the substrate (K_s) becomes large, the rate of proton transfer becomes diffusion controlled. All catalysts then become equally effective independently of their acid strength. Thus the plot of $\log k_A$ versus $\log K_A$ should level off to zero slope when K_A/K_s is large. The converse applies to base catalysis. This means that equations 5.1 and 5.2 can only be approximations, and, although successful over a relatively limited range of acid and base strengths, a curved plot should be obtained over a larger range.

Reactions may show catalysis by basic reagents, although proton transfer is not involved. An example is the catalysis by hydroxide of reactions of esters, where substrate activation by carbonyl addition takes place. This is more usefully termed *nucleophilic catalysis*, and would more reasonably be discussed in Sections 5.2 and 5.3. Several examples are known when reactions following this mechanism can be correlated by equation 5.2. The solvolysis of chloroacetate and bromoacetate (cf. Smith, 1943) are of this type, as is the hydrolysis of esters possessing a good leaving group. Correlation of the hydrolysis of substituted phenyl acetates and ethyl thioacetate catalysed by bases yields β values greater than unity, e.g. 1·6. Clearly hydroxide, the conjugate base of solvent, is not the most effective catalyst. Bender and Turnguest (1957) also find trimethylamine a much weaker catalyst than its basicity would indicate, and imidazole a much more effective one.

5.2 THE SWAIN–SCOTT EQUATION

Largely on the basis of the view that nucleophilic substitution reactions may often also involve a concurrent electrophilic attack on the leaving group,

Swain and Scott (1953) suggested a four-parameter equation

$$\log(k/k_0) = sn + s'e \tag{5.6}$$

for the correlation of polar displacement reactions of all types. In equation 5.6, k/k_0 is the rate of substitution reaction of some substrate relative to some suitable reference substrate; s and s' are substrate parameters; n and e are reagent parameters measuring nucleophilicity and electrophilicity, respectively. The relationship between equation 5.6 and equation 4.19 suggested for solvolysis correlations is evident.

Several special cases of equation 5.6 have already been discussed. Thus when applied to displacements on hydrogen in a series of acids (electrophiles) by a given substrate (constant nucleophile) the Brønsted equation for acid catalysis results

$$\log(k/K^0) = \alpha \log(K_A/K_A^0) = s'(e_A - e_A^0) \tag{5.7}$$

and similarly an acidic substrate (constant electrophile) reacting with a series of bases (nucleophiles) yields

$$\log(k/k^0) = -\beta \log(K_A/K_A^0) = s(n_A - n_A^0) \tag{5.8}$$

For a series of substrates, which are modified by the introduction of substituents, reacting with the same reagent then, for a Hammett series

$$\log(k/k^0) = \sigma(an + be) \tag{5.9}$$

and for a Taft series

$$\log(k/k^0) = \sigma^*(a^*n + b^*e) \tag{5.10}$$

In the special case of the "limiting" or S_N1 solvolysis the substrate has a negligible susceptibility to nucleophilic attack so that,

$$\log(k_A/k_B) = s'(e_A - e_B) = m(\mathbf{Y}_A - \mathbf{Y}_B) \tag{5.11}$$

where the e parameter must now also account for differences in solvent ionization power.

The only example of the use of the full four-parameter equation 5.6 is in the treatment of solvolyses by means of equation 4.19. This has been discussed in Chapter 4.2.

For a given solvent that has substantial electrophilic power, e.g. water, a simpler two-parameter expression has been employed.

$$\log(k/k^0) = sn \tag{5.12}$$

In equation 5.12 k/k^0 is the rate constant for reaction of some substrate with a nucleophile relative to the rate constant for reaction with water, i.e. $n_{H_2O} \equiv 0$. The nucleophilicity parameters are defined by

$$\log(k/k^0)_{CH_3Br} \equiv n \qquad (5.13)$$

using methyl bromide as the standard substrate, i.e. $s \equiv 1$. Similar expressions have been suggested by Hine (1956) and by Wilputte-Steinert and Fierens (1956).

Table 5.3 summarizes the n values, most of which are based on the reactions of the substrates other than methyl bromide. Not all of these reactions were carried out in aqueous solution at 25°C, but the n values do not appear to be very sensitive to temperature nor solvent, provided that this is an aqueous organic medium. The required s values are given in Table 5.4.

TABLE 5.3. NUCLEOPHILIC PARAMETERS IN WATER AND AQUEOUS ORGANIC SOLVENTS
(SWAIN–SCOTT)

Nucleophile	n	Nucleophile	n	Nucleophile	n
1. NO_3^-	1·0	9. $C_6H_5O^-$	3·5	17. SCN^-	4·4[b]
2. Picrate	1·9	10. Br^-	3·5[b]	18. $C_6H_5NH_2$	4·5[a,b]
3. $ClCH_2CO_2^-$	2·2	11. C_5H_5N	3·6	19. I^-	5·0[a,b]
4. $HOCH_2CO_2^-$	2·5	12. HCO_3^-	3·8	20. HS^-	5·1
5. SO_4^{2-}	2·5	13. HPO_4^{2-}	3·8	21. SO_3^{2-}	5·1[b]
6. Cl^-	2·7[b]	14. N_3^-	4·0[a,b]	22. CN^-	5·1
7. $CH_3CO_2^-$	2·7[b]	15. $(NH_2)_2CS$	4·1	23. $S_2O_3^{2-}$	5·35[a,b]
8. HCO_2^-	2·75	16. HO^-	4·2[a,b]	24. $HPSO_3^{2-}$	6·6

a. Defined by equation 5.13. b. Correlates three or more substrates.

The correlations are satisfactory for substitutions at saturated carbon, including the ring-opening reactions of the lactone, the epoxides, and the mustard cation (ethylene-β-chloroethylsulphonium), although in the latter case the reactivity of hydroxide is underestimated by its n value.

The few results for benzoyl chloride are quite well fitted by equation 5.12, although the n value of acetate overestimates its reactivity. The benzene-sulphonyl chloride reactions and the product forming reactions of the trityl cation are poorly correlated. Thus there are clear indications that n values based on substitution at saturated carbon are not always appropriate for substitution at other atoms, e.g. sulphur, or at carbon in other hybridization states. The n values do not parallel basicity, i.e. nucleophilicity towards hydrogen. This is the expected limitation of equation 5.12, and, although its range of application could be intended through the development of other sets of reagent parameters, n_s for reactions at sulphur, etc., the approach described in the next Section appears to be more useful.

TABLE 5.4. SUBSTRATE PARAMETERS (SWAIN–SCOTT)

Substrate	s	Reagents correlated
Methyl bromide	1·00	14; 16; 18; 19; 23.
Methyl iodide	1·15	16; 21; 23.
Methyl p-toluenesulphonate	0·69	14; 16; 17; 23.
Ethyl p-toluenesulphonate	0·66	6; 10; 16; 19.
Benzyl chloride	(0·87)	7; 16.
Benzoyl chloride	1·43	7; 16; 18.
Benzenesulphonyl chloride	(1·25)	16; 18.
Chloroacetate	1·00	3; 7; 8; 12; 16; 21; 23.
Bromoacetate	1·10	3; 4; 7; 8; 16: 23.
Iodoacetate	1·33	9; 16; 17; 22: 23.
β-Propionolactone	0·77	6; 7; 10; 17; 19; 23.
1-Chloro-2,3-epoxypropane	1·00	1; 6; 7; 10; 17; 19.
2,3-Epoxypropanol	0·96	1; 6; 10; 16; 17; 19.
Mustard cation	0·95	2; 5; 6; 7; 11: 12; 13; 15; 17; 18; 19; 20; 21; 23; 24.
Trityl cation	(0·6)	6; 7; 14; 16; 17; 18; 23.

5.3 THE EDWARDS EQUATION

There is a limited correspondence between basicity and nucleophilicity that is responsible for the observation of Brønsted correlations for reactions subject to nucleophilic catalysis. These are usually restricted to a series of reagents all having the same nucleophilic atom, e.g. RCO_2^-. The simplest interpretation of this result is that nucleophilicity can be accounted for by two factors, one of which parallels basicity, while the other, presumably polarizability, remains essentially constant through a series of reagents with a common nucleophilic atom.

Edwards (1954) has sought to account for these two factors by means of equation 5.14

$$\log(k/k_0) = \alpha E_X + \beta H_X \qquad (5.14)$$

where k/k_0 is the rate or equilibrium constant for the reaction of a substrate with a reagent X relative to that for reaction with water; E_X is a parameter measuring the "covalent bonding capacity" of the reagent; α and β are the corresponding substrate parameters. The H_X values are defined by equation 5.15,

$$H_X = pK_{HX} + 1·74 \qquad (5.15)$$

in which $-1·74$, i.e. $-\log 55·5$, is taken as pK of the conjugate acid of the reference reagent, water. The E_X values are defined by equation 5.16

$$E_X = E^0(X_2) + 2 \cdot 60 \qquad (5.16)$$

where $E^0(X_2)$ is the half-cell e.m.f. for the oxidative coupling

$$2X^- \rightarrow X_2 + 2e$$

and the value $-2 \cdot 60 \, V$ is the extrapolated value for $X^- = H_2O$. This suggested measure of nucleophilicity was employed earlier by Foss (1950) but applied only to reactions of sulphur compounds.

Table 5.5 lists the H and E values obtained by means of equations 5.15 and 5.16 and through correlations by equation 5.14. The substrates correlated are summarized in Table 5.6.

TABLE 5.5. REAGENT PARAMETERS (EDWARDS)

Reagent	H	E	P	Reagent	H	E
ClO_4^-	$-9 \cdot 0$[a]	$-0 \cdot 73$		CN^-	$10 \cdot 88$[a]	$2 \cdot 79$
F^-	$4 \cdot 9$[a]	$-0 \cdot 27$	$-0 \cdot 15$	C_5H_5N	$7 \cdot 04$[a]	$1 \cdot 20$
NO_3^-	$0 \cdot 4$	$0 \cdot 29$		$C_6H_5NH_2$	$6 \cdot 28$	$1 \cdot 78$
SO_4^{2-}	$3 \cdot 74$[a]	$0 \cdot 59$[b]		NH_3	$11 \cdot 22$[b]	$1 \cdot 84$[b]
$ClCH_2CO_2^-$	$4 \cdot 54$[a]	$0 \cdot 79$		$(CH_3O)_2POS^-$	$4 \cdot 0$	$2 \cdot 04$[b]
$CH_3CO_2^-$	$6 \cdot 46$[a]	$0 \cdot 95$		SCN	$1 \cdot 0$	$1 \cdot 83$[b]
Cl^-	$-3 \cdot 0$	$1 \cdot 24$[b]	$0 \cdot 39$	$C_7H_7S_2O_3^-$	$-6 \cdot 0$	$2 \cdot 11$[b]
$C_6H_5O^-$	$11 \cdot 74$[a]	$1 \cdot 46$		$C_2H_5S_2O_3^-$	$-5 \cdot 0$	$2 \cdot 06$[b]
Br^-	$-6 \cdot 0$	$1 \cdot 51$[b]	$0 \cdot 54$	$S_2O_3^{2-}$	$3 \cdot 60$[a]	$2 \cdot 52$[b]
N_3^-	$6 \cdot 46$[a]	$1 \cdot 58$		SO_3^{2-}	$9 \cdot 00$[a]	$2 \cdot 57$[b]
HO^-	$17 \cdot 48$[a]	$1 \cdot 65$[b]	$0 \cdot 14$	HS^-	$8 \cdot 70$[a]	$2 \cdot 60$[b]
NO_2^-	$5 \cdot 09$[a]	$1 \cdot 73$[b]		S^{2-}	$14 \cdot 66$[a]	$3 \cdot 08$[b]
CO_3^{2-}	$12 \cdot 1$[a]	$1 \cdot 1$		$(NH_2)_2CS$	$0 \cdot 80$[a]	$2 \cdot 18$
I^-	$-9 \cdot 0$	$2 \cdot 06$[b]	$0 \cdot 72$			

a. H values from equation 5.15. b. E values from equation 5.16.

For all reaction series, except the mutarotation of glucose where a proton transfer rate is involved, the basicity term, βH, contributes in only a minor fashion to $\log(k/k_0)$. This term probably serves mainly as a correlating factor for inadequacies in the E values particularly when these are applied to reactions at a variety of different atoms. Correlations would be somewhat inferior, but still reasonably satisfactory, if the αE term alone were employed. This may seem to be no advance over the Swain–Scott equation. It must be noted on the other hand that the E and H values are "extra-kinetic" parameters obtained from an independent source, whereas the n values are obtained from one member of class of substrates correlated.

TABLE 5.6. SUBSTRATE PARAMETERS (EDWARDS)
(All reactions in aqueous solution at 18–25°C)

Substitution reactions	Site[a]	n[b]	α	β
Ethyl p-toluenesulphonate	C	5	1·68	0·01
β-Propionolactone	C	7	2·00	0·07
Protonated ethylenimine	C	5	2·12	0·03
Diazoacetone	C	4	2·37	0·19
Mustard cation	C	12	2·45	0·07
1-Chloro-2,3-epoxypropane	C	6	2·46	0·04
Methyl bromide	C	6	2·50	0·01
2,3-Epoxypropanol	C	6	2·52	0·00
Iodoacetate	C	6	2·59	−0·05
Benzyl chloride	C	3	3·53	−0·13
Benzoyl chloride	C	4	3·56	0·01
Benzenesulphonyl chloride	S	3	2·56	0·05
Mutarotation of glucose	H	6	−0·41	0·47
H_2O_2 Oxidation	O	5	6·31	−0·39

Complex Formation	n[b]	α	β	Complex Formation	n[b]	α	β
HX^+		0·00	1·00	CuX_2^+	5	7·55	0·28
ZnX^{2+}	6	1·32	0·21	AgX_2^+	15	7·14	−0·25
CuX^{2+}	5	2·04	0·28	HgX_2^{2+}	10	12·92	−0·10
CdX^{2+}	7	2·18	0·07	CuX_4^{2+}	7	4·69	0·96
FeX^{3+}	6	2·52	0·56	ZnX_4^{2+}	11	5·94	0·65
PbX^{2+}	6	2·83	0·27	CdX_4^{2+}	12	6·98	0·26
I_2X	4	3·04	0·00	HgX_4^{2+}	7	16·84	−0·19
AgX^+	8	3·08	−0·08	AuX_4^{3+}	4	26·02	0·29
InX^{3+}	4	3·61	0·34	IX^+	3	6·85	0·16
HgX^{2+}	6	6·57	−0·14				

a. Atom at which substitution takes place. b. Number of reagents other than H_2O correlated.

In a later development of equation 5.14, Edwards (1956) suggests an alternative measure of nucleophilicity that specifically takes into account reagent polarizability. A polarizability parameter, P, is defined by

$$P \equiv \log(R/R_{H_2O}) \qquad (5.17)$$

where R is the molar refraction (see Table 5.6) and used in the modified Edwards equation

$$\log(k/k_0) = aP + bH \qquad (5.18)$$

It is found that equations 5.14 and 5.18 may be transformed into one another by means of

$$a = 3.60\alpha \quad \text{and} \quad b = \beta + 0.62\alpha \tag{5.19}$$

and the E values are given by

$$E = 3.60P + 0.062H \tag{5.20}$$

Equation 5.20 shows that the E values still contain a contribution that parallels basicity and accounts for the, at first sight, surprising result that although the βH term is generally small there are several approximate parallels between nucleophilicity and basicity.

Unfortunately, few P values can be calculated so that equation 5.20 loses much of its interest. Further, Edwards and Pearson (1962) have recognized that theory leads one to expect a polarizability term of the form $R - R_{H_2O}$, unlike the definition given in equation 5.17. It is thus not yet clear whether a meaningful separation if the total, observed relative nucleophilicity has been achieved.

A few discrepancies that need to be clarified include the inability of the E or P value for cyanide adequately to correlate its nucleophilicity and the poor correlations of nucleophilic catalysis of ester reactions. In these reactions, the parallel reagent basicity that yields approximate Brønsted correlations is considerably underestimated by the Edwards parameters. The reactions of protonated ethylenimine are quite well correlated with a small β value, yet a high ρ value (1·5–1·7), implying considerable sensitivity to basicity, is obtained for the Hammett correlations of protonated ethylenimines with substituted benzoates.

An interesting comparison between basicity and oxidative coupling has been discussed by McDaniel and Yingst (1964) in terms of the following thermodynamic cycle. Steps 1 and 4 are the reverse of bond dissociations, steps 2 and 5 are the solvation of neutral molecules, step 3 is acid dissociation and step 6 is the reverse of oxidative coupling.

$$
\begin{array}{ccc}
 & H(g) + X(g) & \\
{}^{1}\nearrow & & \searrow{}^{4} \\
HX(g) & & \tfrac{1}{2}H_2(g) + \tfrac{1}{2}X_2(g) \\
\downarrow {\scriptstyle 2} & & \downarrow {\scriptstyle 5} \\
HX(aq) & & \tfrac{1}{2}(H_2)(aq) + \tfrac{1}{2}X_2(aq) \\
{}^{3}\searrow & H^{+}(aq) + X^{-}(aq) & \nwarrow{}^{6}
\end{array}
$$

Since

$$\Delta H_1 + \Delta H_2 + \Delta H_3 = \Delta H_4 + \Delta H_5 + \Delta H_6 \tag{5.21}$$

it follows, if the solvation terms ΔH_2, ΔH_5 and the terms $T\Delta S_3$, $T\Delta S_6$ are neglected, that

$$1.364 \mathrm{p}K_{HX} - 23.06 E^{0}(X_2) = D_{HX} - \tfrac{1}{2}[D_{H_2} + D_{X_2}] \tag{5.22}$$

(a plot of the left-hand side versus the right-hand side of equation 5.22 is indeed linear with a slope of ~ 1.1).

But the electronegativity may be defined by

$$D_{HX} - \tfrac{1}{2}[D_{H_2} + D_{X_2}] = 23 \cdot 06(x_H - x_X)^2$$

and hence

$$0 \cdot 059 p K_{HX} - E^0(X_2) = (x_H - x_X)^2 \qquad (5.23)$$

A development of equation 5.14 has been described by Davis (1965). For the parameters α_Y and β_Y for some substrate RY, where Y is a variety of leaving groups, it is found, for a fixed R, that α_Y increases linearly with E_Y. Thus reagents that are highly reactive are also good leaving groups. The value of β_Y appears to be related to the electronegativity difference $(x_Y - x_R)^2$ and β_Y increases as Y become more electronegative relative to R.

The values of α listed in the lower part of Table 5.6 provide excellent examples of the concept of "hard" and "soft" acids and bases developed by Pearson (1963). It is suggested the non-polarizable, or "hard", acceptors prefer to form bonds with "hard" donor species, whereas polarizable, or "soft", acceptors prefer "soft" donors. The α values for complex formation become progressively larger, indicating increased response to polarizability as the substrate itself becomes more polarizable. At one extreme, $\alpha = 0$ for the very "hard" acid H^+, and at the other extreme, α is large for the "soft" heavy-metal cations Hg^{2+} and Au^{3+}.

All the correlations so far described refer to aqueous solutions, but may possibly extend to aqueous-solvent mixtures, and perhaps with minor modification to solutions in alcohols. A completely different nucleophilic order is found for solutions in dipolar aprotic solvents (cf. Parker, 1962). Table 5.7 indicates how relative nucleophilicities as measured by the rates of reaction of methyl iodide are completely changed in dimethylformamide solution whereas in methanol solution the customary order is observed.

TABLE 5.7. SOLVENT EFFECT ON NUCLEOPHILICITY

Reagent	k_{rel}(MeOH)	k_{rel}(DMF)	Reagent	k_{rel}(MeOH)	k_{rel}(KMF)
$SeCN^-$	4000	0·16	Br^-	18	0·2
SCN^-	200	0·02	$(CH_3)_2S$	14	0·0001
I^-	100	0·25	Cl^-	1	1
N_3^-	30	0·4	F^-	0·05	>1

The area of LFER's concerned with reagent correlations may well be one of the most fruitful during the next few years. As indicated above considerable empirical success has already been achieved. As studies are extended to include other solvent systems and reaction at other non-metallic sites, such as boron, silicon, nitrogen and phosphorus, it is hoped that the factors controlling reactivity will become better understood.

REFERENCES

Bell, R. P. (1941). "Acid–base Catalysis". Clarendon Press, Oxford.
Bell, R. P. (1959). "The Proton in Chemistry". Methuen, London.
Bender, M. L., and Turnguest, B. W. (1957). *J. Am. chem. Soc.*, **79**, 1656.
Benson, S. W. (1958). *J. Am. chem. Soc.*, **80**, 5151.
Brønsted, J. N., and Pederson, K. J. (1924). *Z. phys. Chem.*, **108**, 185.
Caldin, E. F., and Peacock, J. (1955). *Trans. Faraday Soc.*, **51**, 1217.
Callender, D. D., Ferguson, W. D., Fellis, G. C., and Trotman-Dickinson, A. F. (1960). *J. chem. Soc.*, 3834.
Davis, R. E. (1965). *J. Am. chem. Soc.*, **87**, 3010.
Edwards, J. O. (1954). *J. Am. chem. Soc.*, **76**, 1541.
Edwards, J. O. (1956). *J. Am. chem. Soc.*, **78**, 1819.
Edwards, J. O., Pearson, R. G. (1962). *J. Am. chem. Soc.*, **84**, 16.
Eigen, M. (1964). *Angew. Chem. Internat. Ed.*, **3**, 1.
Foss, O. (1950). *Acta chem. scand.*, **4**, 404, 866.
Gordon, J. E. (1961). *J. org. Chem.*, **26**, 738.
Hine, J. (1956), "Physical Organic Chemistry", pp. 137–138. McGraw-Hill, New York.
McDaniel, D. H., and Yingst, A. (1964). *J. Am. chem. Soc.*, **86**, 1334.
Parker, A. J. (1962). *Q. Rev.*, **16**, 163.
Pearson, R. G. (1963). *J. Am. chem. Soc.*, **85**, 3533.
Smith, G. F. (1943). *J. chem. Soc.*, 521.
Swain, C. G., and Scott, C. B. (1953). *J. Am. chem. Soc.*, **75**, 141.
Wilputte-Steinert, L., and Fierens, P. J. (1956). *Bull. Soc. Chim. belg.*, **65**, 719.

6

Correlation of Spectroscopic Data

Although LFERs, by their very nature, deal only with relative reactivities in the form of reaction-rate and equilibrium data, this approach can be extended to various physical measurements. Any theories of structural and solvent effects that arise from LFER considerations must be expressed ultimately in terms of interactions present within and between molecules. These interactions will influence other quantities, and it is of interest to examine the relationship between the LFER parameters and variations in these quantities. The most outstanding are the various spectroscopic measurements that can in principle be obtained under conditions of greater control, precision and variety than available to reactivity measurements. In addition it is usually possible to relate the observed variation to some theoretically interesting effect more easily than in the case of reactions occurring in solution.

6.1 NUCLEAR MAGNETIC RESONANCE

As soon as a reasonably extensive compilation of proton chemical shifts had been obtained it was inevitable that parallels between these and the various substituent parameters would be sought. Shifts for infinite dilution in inert solvents, such as carbon tetrachloride, may reasonably be taken to be determined largely by intramolecular diamagnetic shielding. In a series of substrates of the type CH_3X, the chemical shift becomes more negative, i.e. resonance occurs at lower applied field strength, as the electronegativity of X increases. However, there are certainly paramagnetic effects arising from the anisotropic magnetic susceptibility of most substituents that obscure any linear relationships between diamagnetic shielding and electronegativity or the σ^* parameters if such relationships do indeed exist. Dailey and Shoolery (1955) have suggested an approximate means of compensation for anisotropic and perhaps other effects, and have examined the *difference* between the methyl and methylene chemical shifts in CH_3CH_2X. An electronegativity correlation on the Pauling scale is given in terms of the relative shifts in parts per million by

$$x_{DS} = 0.695(\delta_{CH_3}^H - \delta_{CH_2}^H) + 1.71 \qquad (6.1)$$

101

However, electronegativities obtained in this way do not always agree with those from other sources (cf. Wells, 1968).

Substituent effects always tend to be better behaved in aromatic than in aliphatic systems, so that more success might be expected in the correlation of proton chemical shifts in substituted benzenes. Indeed Bothner-By and Glick (1956) report a limited correlation of the relative p-proton shifts with the σ_p values. It is clear, however, from an examination of the results of Corio and Dailey (1956), summarized in Table 6.1, that the simple Hammett equation will not yield a satisfactory correlation.

TABLE 6.1. AROMATIC PROTON CHEMICAL SHIFTS
50% v/v solutions in cyclohexane

$$\delta^H = 10^6 \times (H_{C_6H_5X} - H_{C_6H_6})/H_{C_6H_6} \text{ p.p.m.}$$

Substituent	δ_m^H	σ_m	δ_p^H	σ_p^n
$N(CH_3)_2$	0·20	−0·05	0·50	−0·17
NH_2	0·13	−0·04	0·40	−0·17
OCH_3	0·23	0·08	0·23	−0·11
OH	0·37	0·10	0·37	−0·18
CH_3	0·10	−0·07	0·10	−0·13
I	0·17	0·35	0·10	0·30
Br	0	0·39	0	0·27
Cl	0	0·37	0	0·24
CCl_3	−0·17	0·40	−0·23	0·46
CO_2CH_3	−0·20	0·32	−0·27	0·46
$COCH_3$	−0·20	0·38	−0·27	0·50
CN	−0·30	0·61	−0·30	0·67
NO_2	−0·30	0·71	−0·42	0·78

The effect of p-substituents can be reasonably well accounted for in terms of σ_I and σ_R parameters by invoking a larger dependence on resonance effects than is given by the σ_R values. It is difficult, however, to account for the m-substituent effects. The fact that the iodo-group in particular appears quite anomalous suggests that there may be contributions to the shieldings that may be described as "purely magnetic" and bear no simple relationship to the electrostatic and electronic effects measured by the σ parameters.

The effect of substituents on ^{19}F nuclear magnetic shielding in substituted fluorobenzenes is very large and can be quite accurately measured. The studies of Taft et al. (1963) have led to the suggestion that the influence of m-substituents can be accounted for with high precision by means of the σ_I parameters

$$\delta_m^F = -7·10\sigma_I + 0·60 \text{ p.p.m.} \tag{6.2}$$

and the difference between the m- and p-shifts as solely determined by the σ_R^0 parameters

$$\delta_p^F - \delta_m^F = -29 \cdot 5\sigma_R^0 \text{ p.p.m.} \qquad (6.3)$$

These valuable findings have been employed in the compilation of a very extensive list of σ_I and σ_R^0 values (see Taft et al., 1963). The ^{19}F n.m.r. technique can be used under a very wide variety of conditions of solvent to examine substituent effects quite inaccessible to reactivity studies. The effect of solvation, complex formation, acid–base reaction and other factors have been investigated by this means.

The ^{19}F and ^{13}C chemical shifts in aliphatic compounds show, like the proton chemical shifts, some dependence on the electronegativity of substituents, but the situation is quite complicated and as yet not well understood.

In aromatic compounds the ^{19}F and ^{13}C chemical shifts tend to be proportional to one another. In particular Lauterbur (1961) has observed that the difference $\delta_p^c - \delta_m^c$ between the relative p- and m-^{13}C shifts gives an excellent correlation with the σ_R^0 parameter. The m shifts, δ_m^c, cannot be correlated adequately by any of the σ parameters including σ_I. Maciel and Natterstad (1965) consider $\delta_p - \delta_m$ to be a "corrected p shift" dependent upon the π electron density change brought about by the resonance effect of the substituent. The "correction" presumably eliminates any "ring current" effects, inductive effects and perhaps electric field and anisotropic influences.

A direct relationship between π electron density, q, and chemical shift was suggested in the study by Spiesecke and Schneider (1961) of the ^{1}H and ^{13}C spectrum of the aromatic species $C_5H_5^-$, C_6H_6, $C_7H_7^+$ and $C_8H_8^{2+}$. In these cases the π electron densities are 1·20, 1·00, 0·86 and 0·75, respectively, yielding the following

$$\delta^c = 160(q-1); \qquad \delta^H = 10 \cdot 6(q-1) \text{ p.p.m.} \qquad (6.4)$$

Table 6.2 (R. W. Taft, unpublished results) summarizes the calculated π electron density changes, $(q-1)$, and the best estimated value of q at the p-position of the benzene derivative due to resonance interactions as derived from n.m.r. measurements, from the σ_R^0 parameters and from infra-red intensities (see Section 6.3). These estimates were obtained from the following formulae

^{19}F *in substituted fluorobenzenes*

$$(q-1) = (\delta_p^F - \delta_m^F)/200$$

^{13}C *in substituted benzenes*

$$(q-1) = (\delta_p^c - \delta_m^c)/160$$

TABLE 6.2. ESTIMATED π CHARGE DENSITIES IN p-SUBSTITUTED BENZENES $(q-1)$

Substituent	^{19}F	^{13}C	^{13}C'	^{1}H	ν_{16}	$0.15\sigma_R^0$	Best q
O$^-$	0·090	0·097	..	..	..	0·099	1·095
N(CH$_3$)$_2$	0·080	0·080	..	0·085	0·079	0·080	1·080
NH$_2$	0·069	0·068	0·076	0·072	..	0·073	1·071
NHCOCH$_3$	0·029	0·036	..	..	..	0·038	1·034
OCH$_3$	0·063	0·056	0·056	0·056	0·064	0·062	1·060
OH	0·061	0·056	..	..	0·060	0·061	1·059
OCOCH$_3$	0·030	0·023	..	..	0·036	0·032	1·030
F	0·050	0·033	0·050	0·033	0·052	0·053	1·045
Cl	0·026	0·019	0·035	0·017	0·033	0·030	1·026
Br	0·024	0·020	0·035	..	..	0·029	1·025
I	0·020	0·019	0·043	..	..	0·018	1·020
SCH$_3$	0·026	..	..	..	0·038	0·030	1·030
CH$_3$	0·022	0·020	0·025	..	0·018	0·015	1·020
Si(CH$_3$)$_3$	−0·007	0·00	..	..	..	−0·010	0·994
CF$_3$	−0·013	..	..	..	−0·020	−0·020	0·982
CH$_3$CO	−0·027	−0·025	..	..	..	−0·020	0·976
CN	−0·029	..	..	..	−0·017	−0·020	0·978
NO$_2$	−0·028	−0·033	..	−0·020	−0·028	−0·020	0·974

^{13}C' in substituted fluorobenzenes at C—F

$$(q-1) = (\delta_p^{c'} - \delta_m^{c'})/133$$

^{1}H in substituted benzenes

$$(q-1) = (\delta_p^{H} - \delta_m^{H})/6·0$$

ν_{16} i.r. intensity (A) of 1600 cm^{-1} band in substituted benzene

$$(q-1) = A^{\frac{1}{2}}/900$$

6.2 VISIBLE AND U.V. ABSORPTION

Quite profound changes take place in the electronic spectra of aromatic molecules on the introduction of substituents, particularly when resonance interactions can take place. Since these spectra arise from transitions between energy levels associated with the molecule as a whole the differentiation between m- and p-positions no longer has any meaning. The σ_I and σ_R parameters, although believed to be position, and perhaps system, independent, have been obtained from differences between ground-state configurations. No real success has been obtained in attempts to correlate this type of spectroscopic data, possibly because some rather unusual types of interaction are present in the excited states.

On the other hand u.v. and visible spectroscopy has provided much interesting information concerning solvent effects. Kosower (1958) has examined the effect of the medium on the charge-transfer spectra of 1-ethyl-4-carbomethoxypyridinium iodide and several other substrates.

A parameter Z is defined by the molar transition energy (E_T) calculated from the position of absorption maximum ($\tilde{\nu}$ in cm^{-1}) by

$$Z \text{ kcal. mole}^{-1} = 2 \cdot 859 \times 10^{-3}\tilde{\nu} = E_T \tag{6.5}$$

so that high values of Z correspond to large differences between the ground and excited states. These differences must arise largely from solvent stabilization of the ground state, since the time required for an electronic transition is far too short to allow solvent reorganization around the excited state. Some Z values are given in Table 6.3.

TABLE 6.3. SOLVENT Z VALUES

Solvent	Z	Solvent	Z
H_2O	94·6	CH_3CN	71·3
$HOCH_2CH_2OH$	85·1[a]	CH_3SOCH_3	71·1
$H.CONH_2$	83·3	$H.CON(CH_3)_2$	68·5
CH_3OH	83·6	CH_3COCH_3	65·7
C_2H_5OH	79·6	CH_2Cl_2	64·2
$i\text{-}C_3H_7OH$	76·3	C_5H_5N	64·0
$t\text{-}C_4H_9OH$	71·3	$CHCl_3$	63·2
CH_3CO_2H	79·2	C_7H_{16}[b]	60·1

a. Based upon 1-methyl-4-carbomethoxypyridinium iodide. b. Based upon pyridine-N-oxide.

The charge-transfer process responsible for these spectra may be represented by

and it is believed that the charge character of the ion pair ground state and the excited states are such that the resultant molecular dipoles are in perpendicular directions.

This means that highly polar solvent molecules will be orientated about the ground state to give a considerable stabilization, but this orientation will be quite inappropriate for the excited state. An approximate energy diagram illustrating these effects is given below.

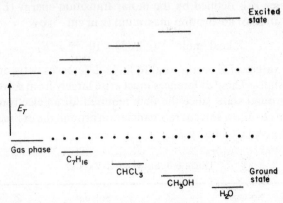

As expected a number of other charge transfer spectra also follow the Z values, thus

For 1-ethyl-4-cyanopyridinium iodide

$$E_T = 0.973Z - 2.08$$

For pyridine-N-oxide

$$E_T = 0.330Z - 81.7$$

For iodide ion in hydroxylic solvents,

$$E_T = -0.203Z + 146.6$$

Many other electronic transitions have pronounced polar character so that, for example, the $n \rightarrow \pi^*$ transition of mesityl oxide follows

$$E_T = 0.155Z + 79.8$$

and the $\pi \rightarrow \pi^*$ transition follows

$$E_T = -0.183Z + 135.1$$

The observation that generated great interest in these solvent parameters is their correlation with the **Y** parameters of Grunwald and Winstein (see Chapter 4.1). The following relationships have been reported

Ethanol–water mixtures

$$\mathbf{Y} = 0.3534Z - 29.45$$

Methanol–water mixtures

$$\mathbf{Y} = 0.4163Z - 35.88$$

Acetone–water mixtures

$$\mathbf{Y} = 0.2989Z - 24.76$$

Dioxan–water mixtures

$$\mathbf{Y} = 0.333Z - 27.58$$

all of which extrapolate to the same $\mathbf{Y}$ and Z values for pure water.

Extension of the Z value treatment to a general correlation of various properties; P, such as reaction rates, equilibria and spectral data in the LFER form

$$P - P_{\text{EtOH}} = SR \qquad (6.6)$$

has been described by Brownstein (1960). The solvent parameters, S, are essentially Kosower's Z values with some additional values from secondary sources. The "reaction" parameters, R, are simply the proportionality factor of the linear relationship 6.6. Good correlations are reported for solvolysis data that follow the $\mathbf{Y}$ values and spectroscopic data, but otherwise rate and equilibrium data and also n.m.r. shifts are rather poorly correlated. It appears that equation 6.6 is restricted to processes closely resembling the defining "reaction series", a comment that can be made concerning the LFER in general.

The whole topic of the empirical correlation of solvent polarity has been reviewed by Reichardt (1965) and various other parameters have been discussed. The theoretical principles underlying current thinking on solvent effects are described.

6.3 I.R. ABSORPTION

Although resultant dipole moments cannot normally be correlated by means of the σ values, individual bond dipoles would presumably be accommodated if they could be accurately and unambiguously determined. I.r. absorption is, however, dependent upon changes in dipole moment during vibrations so that correlations between substituent parameters and i.r. frequency shifts may be expected.

Commonly a correlation between the dimensionless relative frequency and the appropriate σ values is sought in the form of equation 6.7

$$(v - v_0)/v_0 = a\sigma + b \qquad (6.7)$$

Moderately good correlations have been reported for aromatic substrates by Jaffé (1953) as indicated in Table 6.4.

TABLE 6.4. SUBSTITUTED EFFECTS ON I.R. FREQUENCIES

Vibration	Substrate	Solvent	ν_0	a	b	r†
C=O	ArCOCl	CCl$_4$	1778	0·0050	0·0000	0·987
			1744	0·0109	0·0002	0·846
C=O	ArCO$_2$H	CCl$_4$	1743	0·0074	−0·0006	0·973
C=O	ArCO$_2$H	CH$_3$OH	1705	0·0140	0·0000	0·982
C=O	(ArCO$_2$H)$_2$	CCl$_4$	1696	0·0084	0·0000	0·996
C=O	ArCOAr′	CCl$_4$	1664	0·0011	−0·0005	0·955
C=O	ArCOCH$_3$	Nujol	1659	0·0037	0	0·974
NH$_2$	ArNH$_2$	CCl$_4$	3489	0·0098	−0·0003	0·997
			3410	0·0060	−0·0006	0·985
OH	ArOH	CCl$_4$	3551	−0·0030	0·0008	0·948

† r = correlation coefficient.

The C—O stretching frequency in aliphatic alcohols can be correlated with the σ^* values, as can the asymmetric and symmetric stretching frequencies of aliphatic nitro-compounds.

Improvements in methods for determining i.r. intensities have shifted the emphasis towards their correlation, and Brown (1960) has reasoned that the square root of the intensity, $A^{\frac{1}{2}}$, may be the important quantity. For substituted benzophenones and acetophenones, $A^{\frac{1}{2}}$ is found to vary linearly with σ^+, whereas the normal σ values correlate $A^{\frac{1}{2}}$ for substituted phenols, anilines and methylanilines. There are, however, several unexplained deviations.

6.4 MASS SPECTROMETRY

The formation and reactions of ions in the mass spectrometer are processes in which large substituent polar effects must be involved. As yet the LFER has not been extensively applied to mass-spectroscopic results. In this solvent-free environment, much fundamental information will ultimately be forthcoming. This must assist in the understanding of intrinsic substituent effects and, indirectly, of solvent effects.

The ionization potentials and electron affinities obtainable from this type of study are of course important in the determination of electronegativities.

The simplest study of this type so far reported is the ionization potentials of the substituted benzyl radicals reported by Harrison et al. (1961). These are found to be quite well correlated by the σ^+ values, as might have been expected. Continued work along these lines deriving what Kaufmann and Koski (1960) term "absolute Lewis basicities" will be valuable.

REFERENCES

Bothner-By, A. A., and Glick, R. E. (1956). *J. Am. chem. Soc.*, **78**, 1071.
Brown, T. L. (1960). *J. phys. Chem., Ithaca*, **64**, 1798.
Brownstein, S. (1960). *Can. J. Chem.*, **38**, 1590.
Corio, P. L., and Dailey, B. P. (1956). *J. Am. chem. Soc.*, **78**, 3043.
Dailey, B. P., and Shoolery, J. N. (1955). *J. Am. chem. Soc.*, **77**, 3977.
Harrison, A. G., Kebarle, P., and Lossing, F. P. (1961). *J. Am. chem. Soc.*, **83**, 777.
Jaffé, H. H. (1953). *Chem. Rev.*, **53**, 191.
Kaufmann, J. J., and Koski, W. S. (1960). *J. Am. chem. Soc.*, **82**, 3262.
Kosower, E. M. (1958). *J. Am. chem. Soc.*, **80**, 3253, 3261, 3267.
Lauterbur, P. C. (1961). *J. Am. chem. Soc.*, **83**, 1846.
Maciel, G. E., and Natterstad, J. J. (1965). *J. chem. Phys.*, **2**, 2427.
Reichardt, C. (1965). *Angew. Chem. Internat. Ed.*, **4**, 29.
Spiesecke, H., and Schneider, W. G. (1961). *Tetrahedron Lett.*, **14**, 468.
Taft, R. W., Price, E., Fox, I. R., Lewis, I. C., Anderson, K. K., and Davis, G. T. (1963). *J. Am. chem. Soc.*, **88**, 709, 3146.
Wells, P. R. (1968). *Progr. phys. org. Chem.*, **6**, in press.

Author Index

Numbers in *italics* refer to the pages on which the references are listed

Subject Index

A

Acidity functions, 80–88
  H_0, calculations of, 83–85
  H_0, definition of, 80–81
  H_0, table of values, 81, 83, 86
  H_R, definition of, 82
  H_R, table of values, 83
  H_-, definition of, 82
  H_-, table of values, 83
  H_+, definition of, 83
Acid strength
  solvent effects on, 70–80
  substituent effects on, 12, 38, 39
Acrylic acid derivatives, 44
Activation
  enthalpy of, in solvolyses, 62
  entropy of, in solvolyses, 62
  free energy of, correlation of, 1
  free energy of, in solvolyses, 63–65
Activity coefficients, 58
Activity postulate, 70, 81
Adamantane derivatives, 44
Aliphatic substrates
  substituent effects in, observed results, 35–45
  substituent effects in, theoretical treatment, 49–51
Aromatic substrates
  substituent effects in, observed results, 8–35
  substituent effects in, theoretical treatment, 51–56
α Values
  See Relay factors, Coulombic integral and Substrate parameters
  for *m*-resonance effects, 17

B

Bicyclo-octane derivatives, 43, 44
Brønsted equation, 2, 89–92
t-Butyl chloride, relative solvolysis rates, 60–66

β Values

β Values
  See Resonance integrals and Substrate parameters (Brønsted and Edwards)
  Isokinetic temperature, 21–23

C

Catalysis
  acid, dependence on acidity functions, 86–88
  general acid, correlation of, 89–92
  general base, correlation of, 89–92
  nucleophilic, 92
  specific acid, 89
  specific base, 89
Conjugative interactions, 10
Correlation coefficient, r, 3
Coulombic integral, 52
Crotonic acid derivatives, 44
Cyclohexane derivatives, 43, 44

D

Direct electrostatic effects, 10, 43, 44
  in aliphatic substrates, 50, 51
  in aromatic substrates, 51, 52
Direct resonance effects, 10, 27

E

E Values, *See* Reagent parameter (Edwards)
E_s Values, steric-effect parameter, 41, 42
Enthalpy changes
  internal and external contributions, 22, 23, 48
  of activation, in solvolyses, 62
Entropy changes
  internal and external contributions, 22, 23, 48
  of activation, in solvolyses, 62
Equilibria, generalized form, 47
Ester reactions, substituent effects on, 12, 35, 36
"External" enthalpy and entropy changes, 22, 23, 48